DRIFT INTO FAILURE

DRIFT INTO FAILURE

From Hunting Broken Components to Understanding Complex Systems

Sidney Dekker

CRC Press
Taylor & Francis Group
Boca Raton London New York

CRC Press is an imprint of the
Taylor & Francis Group, an **informa** business

CRC Press
Taylor & Francis Group
6000 Broken Sound Parkway NW, Suite 300
Boca Raton, FL 33487-2742

© 2011 by Sidney Dekker
CRC Press is an imprint of Taylor & Francis Group, an Informa business

No claim to original U.S. Government works

Printed on acid-free paper
Version Date: 20160226

International Standard Book Number-13: 978-1-4094-2222-8 (Hardback) 978-1-4094-2221-1 (Paperback)

Visit the Taylor & Francis Web site at
http://www.taylorandfrancis.com

and the CRC Press Web site at
http://www.crcpress.com

CONTENTS

LIST OF FIGURES

ACKNOWLEDGMENTS

I want to thank Paul Cilliers and Jannie Hofmeyr at the Centre for Studies in Complexity at the University of Stellenbosch, South Africa, for our fascinating discussions on complexity, ethics and system failure. I also want to thank Eric Wahren and Darrell Horn for their studious reading of earlier drafts and their helpful comments for improvement.

Reviews for
Drift into Failure

'"*Accidents come from relationships, not broken parts.*" *Sidney Dekker's meticulously researched and engagingly written* Drift into Failure: From Hunting Broken Parts to Understanding Complex Systems *explains complex system failures and offers practical recommendations for their investigation and prevention from the combined perspectives of unruly technology, complexity theory, and post-Newtonian analysis. A valuable source book for anyone responsible for, or interested in, organizational safety.*'

Steven P. Bezman, Aviation Safety Researcher

'*Dekker's book challenges the current prevalent notions about accident causation and system safety. He argues that even now, what profess to be systemic approaches to explaining accidents are still caught within a limited framework of 'cause and effect' thinking, with its origins in the work of Descartes and Newton. Instead, Dekker draws his inspiration from the science of complexity and theorises how seemingly reasonable actions at a local level may promulgate and proliferate in unseen (and unknowable) ways until finally some apparent system "failure" occurs. The book is liberally illustrated with detailed case studies to articulate these ideas. As with all Dekker's books, the text walks a fine line between making a persuasive argument and provoking an argument. Love it or hate it, you can't ignore it.*'

Don Harris, HFI Solutions Ltd

'*Dekker's book contributes to the growing debate around the nature of retrospective investigations of safety-critical situations in complex systems. Both provocative and insightful, the author shines a powerful light on the severe limits of traditional linear approaches. His call for a diversity of voices and narratives, to deepen our understanding of accidents, will be welcomed in healthcare. Dekker's proposal that we shift from going "down and in" to "up and out" suggests a paradigm shift in accident investigation.*'

Rob Robson, Healthcare System Safety and Accountability, Canada

'*Professor Dekker explodes the myth that complex economic, technological and environmental failures can be investigated by approaches fossilized in linear, Newtonian-Cartesian logic. Today nearly 7 billion people unconsciously reshape themselves, their organizations, and societies through the use of rapidly-evolving, proliferating and miniaturizing technologies powered by programs that supersede the intellectual grasp of their developers. Serious proponents of the next high reliability organizations would do well to absorb* Drift into Failure.'

Jerry Poje, Founding Board Member of the U.S. Chemical Safety and Hazard
Investigation Board

'*Today, catastrophic accidents resulting from failure of simple components confound industry. In* Drift into Failure, *Dekker shows how reductionist analysis – breaking the system down until we find the "broken part" – does not explain why accidents in complex systems occur. Dekker introduces the systems approach. Reductionism delivers an inventory of broken parts; Dekker's book offers a genuine possibility of future prevention. The systems approach may allow us to Drift into Success.*'

John O'Meara, HAZOZ

PREFACE

When I was in graduate school for my doctorate, we always talked about the systems we studied as *complex* and *dynamic*. Aviation, nuclear power, medicine, process control – these were the industries that we were interested in, and that seemed to defy simple, linear modeling – industries that demand of us, researchers, safety analysts, a commitment to penetrate the elaborate, intricate and live ways in which their work ebbs and flows, in which human expertise is applied, how organizational, economic and political forces suffuse and constrain their functioning over time.

Back then, and during most of my work in the years since, I have not encountered many models that are complex or dynamic. Instead, they are mostly simple and static. Granted, models are models for a reason: they are abstractions, simplifications, or perhaps no more than hopes, projections. Were a perfect model possible, one that completely and accurately represented the dynamics and complexity of its object, then its very specificity would defeat the purpose of modeling.

So models always make sacrifices of some kind. The question, though, is whether our models sacrifice inconsequential aspects of the worlds we wish to understand and control, or vital aspects.

During my first quarter in graduate school I took five classes, thinking this would be no problem. Well, actually, I didn't really think about it much at all. What concerned me was that I wanted as much value for money as I could get. I paid for my first quarter in grad school myself, which, for an international student, was a significant outlay (from there on I became a Graduate Research Assistant and the tuition was waived, otherwise I would not be writing this or much of anything else). For that first quarter, the earnings from a summer consulting job were burnt in one invoice. Then something interesting happened. Somehow I found out that with the four classes I *had* to take, I had reached a kind of maximum level beyond which apparently not even Ohio State could morally muster to extort more money from its international students. I could, in other words, throw in a class for the fun of it.

I did.

It became a class in non-linear dynamic systems. The choice was whimsical, really, a hint from a fellow student, and a fascinating title that seemed to echo some of the central labels of the field I was about to pursue a PhD in. The class was taught at the Department of Psychology, mind you. The room was small and dark and dingy and five or six students sat huddled around the professor.

The first class hit me like the blast of a jet engine.

The differences between static and dynamic stability were the easy stuff. You know, like done and over with in the first three minutes. From there, the professor galloped through an increasingly abstruse, dizzying computational landscape of the measurement of unpredictability, rotating cylinders and turning points, turbulence and dripping faucets, strange attractors, loops in phase space, transitions, jagged shores and fractals. And the snowflake puzzle.

I didn't get an A.

It was not long after James Gleick had published *Chaos*, and a popular fascination with the new science of complexity was brewing. The same year that I took this class, 1992, Roger Lewin published the first edition of *Complexity*, a first-person account of the adventures of people at the Santa Fe Institute and other exciting places of research.

Taking this class was in a sense a fractal, a feature of a complex system that can be reproduced at any scale, any resolution. The class talked about the butterfly effect, but it also set in motion a butterfly effect. In one sense, the class was a marginal, serendipitous footnote to the subsequent years in grad school. But it represented the slightest shift in starting conditions. A shift that I wouldn't have experienced if it hadn't been for the tuition rules (or if I hadn't been reminded of that particularly arcane corner of the tuition rules or met that particular student who suggested the class, or if the psychology department hadn't had a professor infatuated with computation and complexity), an infinitesimal change in starting conditions that might have enormous consequences later on.

Well, if you consider the publication of yet another book "enormous." Hardly, I agree. But still, I was forced to try to wrap my arms around the idea that complex, dynamic systems reveal adaptive behavior more akin to living organisms than the machines to which most safety models seem wedded. By doing this, the seeds of complexity and systems thinking were planted in me some 20 years ago.

Drifting into failure is a gradual, incremental decline into disaster driven by environmental pressure, unruly technology and social processes that normalize growing risk. No organization is exempt from drifting into failure. The reason is that routes to failure trace through the structures, processes and tasks that are necessary to make an organization successful. Failure does not come from the occasional, abnormal dysfunction or breakdown of these structures, processes and tasks, but is an inevitable by-product of their normal functioning. The same characteristics that guarantee the fulfillment of the organization's mandate will turn out to be responsible for undermining that mandate.

Drifting into failure is a slow, incremental process. An organization, using all its resources in pursuit of its mandate (providing safe air-travel, delivering electricity reliably, taking care of your savings), gradually borrows more and more from the margins that once buffered it from assumed boundaries of failure. The very pursuit of the mandate, over time, and under the pressure of various environmental factors (competition and scarcity most prominently), dictates that it does this borrowing – does things more efficiently, does more with less, perhaps takes greater risks. Thus, it is the very pursuit of the mandate that creates the conditions for its eventual collapse. The bright side inexorably brews the dark side – given enough time, enough uncertainty, enough pressure. The empirical base is not very forgiving: Even well-run organizations exhibit this pattern.

This reading of how organizations fail contradicts traditional, and some would say simplistic, ideas about how component failures are necessary to explain accidents. The traditional model would claim that for accidents to happen, something must break, something must give, something must malfunction. This may be a component part, or a person. But in stories of drift into failure, organizations fail precisely because they are doing well – on a narrow range of performance criteria, that is – the ones that they get rewarded on in their current political or economic or commercial configuration. In the drift into failure, accidents can happen without anything breaking, without anybody erring, without anybody violating the rules they consider relevant.

I believe that our conceptual apparatus for understanding drift into failure is not yet well-developed. In fact, most of our understanding is held hostage by a Newtonian–Cartesian vision of how the world works. This makes particular (and often entirely taken-for-granted) assumptions about decomposability and the relationship between cause and effect. These assumptions may be appropriate for understanding simpler systems, but are becoming increasingly inadequate for examining how formal-bureaucratically organized risk management, in a tightly interconnected complex world, contributes to the incubation of failure.

The growth of complexity in society has outpaced our understanding of how complex systems work and fail. Our technologies have got ahead of our theories. We are able to build things whose properties we understand in isolation. But in competitive, regulated societies, their connections proliferate, their interactions and interdependencies multiply, their complexities mushroom.

In this book, I explore complexity theory and systems thinking to better understand how complex systems drift into failure. I take some of the ideas from that early class in complexity theory, like sensitive dependence on initial conditions, unruly technology, tipping points, diversity – to find that failure emerges opportunistically, non-randomly, from the very webs of relationships that breed success and that are supposed to protect organizations from disaster. I hope this book will help us develop a vocabulary that allows us to harness complexity and find new ways of managing drift.

1

FAILURE IS ALWAYS AN OPTION

Accidents are the effect of a systematic migration of organizational behavior under the influence of pressure toward cost-effectiveness in an aggressive, competitive environment.[1]

Rasmussen and Svedung

WHO MESSED UP HERE?

If only there was an easy, unequivocal answer to that question. In June 2010, the U.S. Geological Survey calculated that as much as 50,000 barrels, or 2.1 million gallons of oil a day, were flowing into the Gulf of Mexico out of the well left over from a sunken oil platform. The Deepwater Horizon oil rig exploded in April 2010, then sank to the bottom of the sea while killing 11 people. It triggered a spill that lasted for months as its severed riser pipe kept spewing oil deep into the sea.

Anger over the deaths and unprecedented ecological destruction turned to a hunt for culprits – Tony Hayward, the British CEO of BP, which used the rig (the rig was run by Transocean, a smaller exploration company) or Carl-Henric Svanberg, its Swedish chairman, or people at the federal Minerals Management Service. As we wade deeper into the mess of accidents like these, the story quickly grows murkier, branching out into multiple possible versions. The "accidental" seems to become less obvious, and the roles of human agency, decision-making and organizational trade-offs appear to grow in importance. But the possible interpretations of *why* these decisions and trade-offs caused an oil rig to blow up are book-ended by two dramatically different families of versions of the story.

Ultimately, these families of explanations have their roots in entirely different assumptions about the nature of knowledge (and, by extension, human decision-making). These families present different premises about how events are related to each other through cause and effect, and about the foreseeability and preventability of disasters and other outcomes. In short, they take very different views of how the world can be known, how the world works, and how it can be controlled or influenced. These assumptions tacitly inform much of what either family sees as common-sense: which stones it should look for and turn over to find the sources of disaster. When we respond to failure, we may not even know that we are firmly in-family in one way or another. It seems so natural, so obvious, so taken-for-granted to ask the questions we ask, to look for causes in the places we do.

One family of explanations goes back to how the entire petroleum industry is rotten to the core, how it is run by callous men and not controlled by toothless regulators and corruptible governments. More powerful than many of the states in which its operates, the industry has governments in its pocket. Managers spend their days making amoral trade-offs to the detriment of nature and humanity. Worker safety gets sacrificed, as do environmental concerns, all in the single-minded and greedy pursuit of ever greater profits.[2] Certain managers are more ruthless than others, certain regulators more hapless than others, some workers more willing to cut corners than others, and certain governments easier to buy than others. But that is where the differences essentially end. The central, common problem is one of culprits, driven by production, expediency and profit, and their unethical decisions. Fines and criminal trials will deal with them. Or at least they will make us feel better.

The family of explanations that identifies bad causes (bad people, bad decisions, broken parts) for bad outcomes is firmly quartered in the epistemological space[3] once established by titans of the scientific revolution – Isaac Newton (1642–1727) and René Descartes (1596–1650). The model itself is founded in, and constantly nourished by, a vision of how the world works that is at least three centuries old, and which we have equated with "analytic" and "scientific" and "rational" ever since. In this book, I call it the Newtonian–Cartesian vision.[4]

Nowadays, this epistemological space is populated by theories that faithfully reproduce Cartesian and Newtonian ideas, and that make us think about failure in their terms. We might not even be aware of it, and, more problematically, we might even call these theories "systemic." Thinking about risk in terms of energy-to-be-contained, which requires barriers or layers of defense, is one of those faithful reproductions. The linear sequence of events (of causes and effects) that breaks through these barriers is another. The belief that, by applying the right method or the best method, we can approximate the true story of what happened is Newtonian too: it assumes that there is a final, most accurate description of the world. And underneath all of this, of course, is a reproduction of the strongest Newtonian commitment of all: reductionism. If you want to understand how something works or fails, you have to take it apart and look at the functioning or non-functioning of the parts inside it (for example, holes in a layer of defense). That will explain why the whole failed or worked.

RATIONAL CHOICE THEORY

The Newtonian vision has had enormous consequences for our thinking even in the case of systems that are not as linear and closed as Newton's basic model – the planetary system. Human decision-making and its role in the creation of failure and success is one area where Newtonian thought appears very strongly. For its psychological and moral nourishment, this family of explanations runs on a variant of rational choice theory. In the words of Scott Page:

> In the literature on institutions, rational choice has become the benchmark behavioral assumption. Individuals, parties, and firms are assumed to take actions that optimize their utilities conditional on their information and the actions of others. This is not inconsistent with that fact that, ex post, many actions appear to be far from optimal.[5]

Rational choice theory says that operators and managers and other people in organizations make decisions by systematically and consciously weighing all possible outcomes along all relevant criteria. They know that failure is always an option, but the costs and benefits of decision alternatives that make such failure more or less likely are worked out and listed. Then people make a decision based on the outcome that provides the highest utility, or the highest return on the criteria that matter most, the greatest benefit for the least cost. If decisions after the fact ("ex post" as Scott Page calls it) don't seem to be optimal, then something was wrong with how people inside organizations gathered and weighed information. They should or could have tried harder. BP, for example, hardly seems to have achieved an optimum in any utilitarian terms with its decision to skimp on safety systems and adequate blowout protection in its deepwater oil pumping. A few more million dollars in investment here and there (a couple of hours of earnings, really) pretty much pales in comparison to the billions in claims, drop in share price, consumer boycotts and the immeasurable cost in reputation it suffered instead – not to mention the 11 dead workers and destroyed eco-systems that will affect people way beyond BP or its future survival.

The rational decision-maker, when she or he achieves the optimum, meets a number of criteria. The first is that the decision-maker is completely informed: she or he knows all the possible alternatives and knows which courses of action will lead to which alternative. The decision-maker is also capable of an objective, logical analysis of all available evidence on what would constitute the smartest alternative, and is capable of seeing the finest differences between choice alternatives. Finally, the decision-maker is fully rational and able to rank the alternatives according to their utility relative to the goals the decision-maker finds important. These criteria were once formalized in what was called Subjective Expected Utility Theory. It was devised by economists and mathematicians to explain (and even guide) human decision-making. Its four basic assumptions were that people have a clearly defined utility function that allows them to index alternatives according to their desirability, that they have an exhaustive view of decision alternatives, that they can foresee the probability of each alternative scenario and that they can choose among those to achieve the highest subjective utility.

A strong case can be made that BP should have known all of this, and thus should have known better. U.S. House Representative Henry Waxman, whose Energy and Commerce Committee had searched 30,000 BP documents looking for evidence of attention to the risks of the Deepwater well, told the BP chairman,

"There is not a single email or document that shows you paid even the slightest attention to the dangers at the well. You cut corner after corner to save a million dollars here and a few hours there. And now the whole Gulf Coast is paying the price."[6] This sounded like amoral calculation – of willingly, consciously putting production before safety, of making a deliberate, rational calculation of rewards and drawbacks and deciding for saving money and against investing in safety.

And it wasn't as if there was no precedent to interpret BP actions in those terms. There was a felony conviction after an illegal waste-dumping in Alaska in 1999, criminal convictions after the 2005 refinery blast that killed 15 people in Texas City, and criminal convictions after a 2006 Prudhoe Bay pipeline spill that released some 200,000 gallons of oil onto the North Slope. After the 2005 Texas City explosion, an independent expert committee concluded that "significant process safety issues exist at all five U.S. refineries, not just Texas City," and that "instances of a lack of operating discipline, toleration of serious deviations from safe operating practices, and apparent complacency toward serious process safety risk existed at each refinery."[7] The panel had identified systemic problems in the maintenance and inspection of various BP sites, and found a disconnect between management's stated commitment to safety and what it actually was willing to invest. Unacceptable maintenance backlogs had ballooned in Alaska and elsewhere. BP had to get serious about addressing the underlying integrity issues, otherwise any other action would only have a very limited or temporary effect.

It could all be read as amoral calculation. In fact, that's what the report came up with: "Many of the people interviewed … felt pressured to put production ahead of safety and quality."[8] The panel concluded that BP had neglected to clean and check pressure valves, emergency shutoff valves, automatic emergency shutdown mechanisms and gas and fire safety detection devices (something that would show up in the Gulf of Mexico explosion again), all of them essential to preventing a major explosion. It warned management of the need to update those systems, because of their immediate safety or environmental impact. Yet workers who came forward with concerns about safety were sanctioned (even fired in one case), which quickly shut down the flow of safety-related information.

Even before getting the BP chairman to testify, the U.S. congress weighed in with its interpretation that bad rational choices were made, saying "it appears that BP repeatedly chose risky procedures in order to reduce costs and save time, and made minimal efforts to contain the added risk." Many people expressed later that they felt pressure from BP to save costs where they could, particularly on maintenance and testing. Even contractors received a 25 percent bonus tied to BP's production numbers, which sent a pretty clear message about where the priorities lay. Contractors were discouraged from reporting high occupational health and safety statistics too, as this would ultimately interfere with production.[9]

Rational choice theory is an essentially economic model of decision-making that keeps percolating into our understanding of how people and organizations work and mess up. Despite findings in psychology and sociology that deny that people have the capacity to work in a fully rational way, it is so pervasive and so subtle that we might hardly notice it. It affects where we look for the causes of disaster (in people's bad decisions or other broken parts). And it affects how we assess the morality of, and accountability for, those decisions. We can expect people involved in a safety-critical activity to know its risks, to know possible outcomes, or to at least do their best to achieve as great a level of knowledge about it as possible. What it takes on their part is an effort to understand those risks and possible outcomes, to plot them out. And it takes a moral commitment to avoid the worst of them. If people knew in advance what the benefits and costs of particular decision alternatives were, but went ahead anyway, then we can call them amoral.

The amoral calculator idea has been at the head of the most common family of explanations of failure ever since the early 1970s. During that time, in response to large and high-visibility disasters (Tenerife, Three Mile Island), a historical shift occurred in how societies understood accidents.[10] Rather than as acts of God, or fate, or meaningless (that is, truly "accidental") coincidences of space and time, accidents began to be seen as failures of risk management. Increasingly, accidents were constructed as human failures, as organizational failures. As moral failures.

The idea of the amoral calculator, of course, works only if we can prove that people knew, or could reasonably have known, that things were going to go wrong as a result of their decisions. Since the 1970s, we have "proven" this time and again in accident inquiries (for which the public costs have risen sharply since the 1970s) and courts of law. Our conclusions are most often that bad or miscreant people made amoral trade-offs, that they didn't invest enough effort, or that they were negligent in their understanding of how their own system worked. Such findings not only instantiate, but keep reproducing the Newtonian–Cartesian logic that is so common-sense to us. We hardly see it anymore, it has become almost transparent. Our activities in the wake of failure are steeped in the language of this worldview. Accident inquiries are supposed to return probable "causes." The people who participate in them are expected by media and industry to explain themselves and their work in terms of broken parts (we have found what was wrong: here it is). Even so-called "systemic" accident models serve as a vehicle to find broken parts, though higher upstream, away from the sharp end (deficient supervision, insufficient leadership). In courts, we argue that people could reasonably have foreseen harm, and that harm was indeed "caused" by their action or omission. We couple assessments of the extent of negligence, or the depth of the moral depravity of people's decisions, to the size of the outcome. If the outcome was worse (more oil leakage, more dead bodies), then the actions that led up to it must have been really, really bad. The fine gets higher, the prison sentence longer.

It is not, of course, that applying this family of explanations leads to results that are simply false. That would be an unsustainable and useless position to take. If the worldview behind these explanations remains invisible to us, however, we will never be able to discover just how it influences our own rationalities. We will not be able to question it, nor our own assumptions. We might simply assume that this is the only way to look at the world. And *that* is a severe restriction, a restriction that matters. Applying this worldview, after all, leads to particular results. It doesn't really allow us to escape the epistemological space established more than 300 years ago. And because of that, it necessarily excludes other readings and other results. By not considering those (and not even knowing that we *can* consider those alternatives) we may well short-change ourselves. It may leave us less diverse, less able to respond in novel or more useful ways. And it could be that disasters repeat themselves because of that.

TECHNOLOGY HAS DEVELOPED MORE QUICKLY THAN THEORY

The message of this book is simple. The growth of complexity in society has got ahead of our understanding of how complex systems work and fail. Our technologies have gone ahead of our theories.[11] We are able to build things – from deep-sea oil rigs to jackscrews to collaterized debt obligations – whose properties we can model and understand *in isolation*. But, when released into competitive, nominally regulated societies, their connections proliferate, their interactions and interdependencies multiply, their complexities mushroom. And we are caught short.

We have no well-developed theories for understanding how such complexity develops. And when such complexity fails, we still apply simple, linear, componential ideas as if those will help us understand what went wrong. This book will argue that they won't, and that they never will. Complexity is a defining characteristic of society and many of its technologies today. Yet simplicity and linearity remain the defining characteristics of the theories we use to explain bad events that emerge from this complexity. Our language and logic remain imprisoned in the space of linear interactions and component failures that was once defined by Newton and Descartes.

When we see the negative effects of the mushrooming complexity of our highly interdependent society today (an oil leak, a plane crash, a global financial crisis), we are often confident that we can figure out what went wrong – if only we can get our hands on *the part that broke* (which is often synonymous to getting our hands on the human(s) who messed up). Newton, after all, told us that for every effect there is an equal and opposite cause. So we can set out and trace back from the foreclosed home, the smoking hole in the ground or the oil-spewing hole in the sea floor, and find that cause. Analyses of breakdowns in complex systems remain depressingly linear, depressingly componential.

This doesn't work only when we are faced with the rubble of an oil rig, or a financial crisis or an intractable sovereign debt problem. When we put such technologies to work, and regulate them, we may be overconfident that we can foresee the effects, because we apply Newtonian folk-science to our understanding of how the world works. With this, we make risk assessments and calculate failure probabilities. But in complex systems, we can never predict results, we can only indicate them and their possibility. We can safely say that some mortgage lenders will get into trouble, that some people will lose their houses in foreclosure, that there will be an oil leak somewhere, or a plane crash. But who, what, where, and when? Only a Newtonian universe allows such precision in prediction. We don't live in a Newtonian universe any longer – if we ever did.

But if we want to understand the failings of complex systems, whether before or after, we should not put too much confidence in theories that were developed on a philosophy for simple, closed, linear systems. We have to stop just relying on theories that have their bases in commitments about knowledge, about the world, and about the role of science and analysis that are more than three centuries old. We have to stop just relying on theories that take as their input data only the synchronic snapshot of how the system lies in pieces when we find it (yes, "ex post") – when we encounter it broken, with perforated layers of defense. These theories and philosophical commitments have their place, their use, and their usefulness. But explaining complexity may not be one of them.

Remember the message of this book: the complexity of what society and commerce can give rise to today is not matched by the theories we have that can explain why such things go wrong. If we want to understand the failings of complexity, we have to engage with theory that can illuminate complexity. Fortunately, we have pretty solid and exciting bases for such theories today. What is complexity? Why is it so different, and so immune against the approaches of simplifying, reducing, of drawing straight lines between cause and effect, chopping up, going down and in? Why does it want to reject logics of action and intervention that once upon a time worked so well for us? Some of the answers lie, as you might expect, in complexity theory. Or, as it is also known, in complexity and systems theory. Or in the theory of complex adaptive systems. The label matters less than the usefulness of what such a theory can tell us. That is what this book sets out to do in its latter half: delve into complexity theory, mine it for what it is worth, discover what it can tell us about how complex systems work and fail. And what we can (and cannot) do about it.

Systems with only a few components and few interdependencies are not going to generate complexity. Complexity means that a huge number of interacting and diverse parts give rise to outcomes that are really hard, if not impossible, to foresee. The parts are connected, interacting, diverse, and together they generate adaptive behavior in interaction with their environment. It is their interdependencies and interactions that is responsible for their ability to produce adaptation. Complex systems pulse with life. The way in which components are interlinked, related and cross-adaptive, or interactive despite and because their diversity, can give rise to novelty, to large events. These effects that are typically emergent, that is, they are impossible to locate back in the properties or micro-level behavior of any one of the components. Higher-level or macro-level structures and patterns and events are the joint product of the behavior of complex systems. But which structures and patterns might emerge, and how, is rather unpredictable. Today, more than ever before, complex systems are everywhere.

THE GAUSSIAN COPULA

One of the beautiful things about complexity is that we couldn't design it, even if we wanted to. If something so complex could be designed, it wouldn't be complex, because it would all have to fit inside the head or computer model of a designer or a design team. There are hard computational limits, as well as limits of knowledge and foresight, which make that a designable system actually has to be simple (or merely complicated, the difference will be explained later in the book). Complexity is not designed. Complexity happens. Complexity grows, and it grows on itself, it typically grows more of itself. This doesn't mean that we can't design parts of a complex system, or design something that will be usurped in webs of interactions and interdependencies. This is what happened with the Gaussian copula. Designed in isolation, it was a wonderful thing for the people who used it to assess risk and make money. As more and more webs of interactions and relationships and interdependencies and feedback loops started growing around it, however, it became part of a complex system. And as such, it became really hard to foresee how it could bring global lending to a virtual standstill, triggering a worldwide financial crisis and a deep recession. Once the copula was an enabler of mortgaging the most hopeless of homeowner prospects. Now it was the trigger of a recession that swelled the number of homeless families in the U.S. by 30 percent inside two years.[12]

The Gaussian copula (1) was an equation intended to price collaterized debt obligations. It was a function concocted by a mathematician called David Li, who was trying to look at lots of different bonds (particularly collaterized debt obligations, more about those in a second) and work out whether they were moving in the same direction or not. The Gaussian, of course, is about probabilities, and whether they are associated with other probabilities. Basically, by putting in lots of different bonds the Gaussian copula function produced a single number that became easily manipulable and trackable by the world of quantitative finance. It could show correlations between bonds that might default and bonds that might not. The financial world loved it. Here was one number, coughed up by a relatively simple equation, with which they could trade a million different securities (a security is simply something that shows ownership, or right of ownership of stocks or bonds, or the right to ownership connected with derivatives that in turn get their value from some underlying asset).

$$\Pr[T_A<1,T_B<1] = \Phi_2(\Phi^{-1}(F_A(1)), \Phi^{-1}(F_B(1)),\gamma) \tag{1}$$

In the Gaussian copula, probability Pr is a joint default probability, the likelihood that any two members of a pool (A and B, each of which might contain hundreds or even thousands of mortgages, for example) will both default. This is what investors are looking for, and the rest of the copula provides the answer. Φ_2 is the copula: it couples (hence the Latin *copula*) the individual probabilities associated with pools A and B to come up with a single number. Errors here can massively increase the risk of the whole equation producing garbage. Survival times (T_A and T_B) represent the amount of time between now and when A and B can be expected to default (that is, fail to repay). Li was inspired in this by actuarial science, which has a similar concept for what happens to somebody's life expectancy when their spouse dies. Distribution functions F_A and F_B represent the probabilities of how long A and B are likely to survive. These are not certainties, so small misassessments or miscalculations here can lead to a much greater production or risk than the formula would have you believe. The idea about equality (=) between the copula and probability of default might be a dangerously misleading concept in this formula, as it suggests that there is no room for error, and is a short notation that says "is" or "equals" or "is equal to," which muffles the role of real-world uncertainty, fuzziness, and precariousness. The gamma γ at the end is the all-powerful correlation parameter, which reduces correlation to a single constant. This, in this context, is something that should be highly improbable, if not impossible. It was, however, the part that really made Li's idea irresistible and thereby so pervasive.

In isolation it looked elegant, and it worked even more elegantly. But what was the world in which the Gaussian copula was released? It was the world of collaterized debt obligations, which was growing exponentially in size and complexity from the 1990s onward. Collaterized debt obligations, an invention from the late 1980s, are a type of synthetic asset-backed security. Some sort of corporate entity needs to be constructed to hold assets as collateral (say, somebody's house), and then collect interest which can be sold as packages of cash flow to investors. Collaterized debt obligations can come either from a special purpose entity that acquires a portfolio of underlying assets such as mortgage-backed securities, commercial real estate bonds and corporate loans, or from the special purpose entity issues bonds (those collaterized debt obligations) in what are called different tranches (levels of risk), from which the proceeds are then used to purchase a portfolio of underlying assets.

The risk and return for somebody who has invested in a debt obligation (often without really knowing it, by the way) depends on how the obligations and their tranches are defined (and this may not be communicated very clearly or get lost in the various layers of buying, selling and reselling), and only indirectly on the underlying assets. So actually, the investment depends on the assumptions and methods used to define the risk and return of the tranches. Like all asset-backed securities, debt obligations enable the originators of the underlying assets to pass credit risk to a variety of other institutions or chop it up and distribute it to an immeasurable number of individual investors, who in turn might resell it. Risk is distributed and sold into invisibility. With the multiplying layers of players involved in the bond market in sending this money (or debts) around, risks were chopped up and scattered into oblivion.

The use and increasing opacity of financial instruments like debt obligations expanded dramatically on the back of a growing number of asset managers and investors. There was no intelligent design behind this, no single smart designer who had it all figured out beforehand, or could figure it all out even while things were playing out. The growth of asset management was in part a response to – and source for – a growing need for stock market investments and mortgages. From the 1990s onward, an increasing set of players in modern society (from private individuals to government pension funds to sovereign nations) turned to stock markets for financial returns and presumed future security. It intersected with the realization in many developed nations of a demographic time bomb that would leave them with way too many pensioners and way too few people doing productive taxable work. Pension funds were going to be depleted if nothing was done, and many countries embarked on aggressive strategies to get people to invest in their own future pensions. At the same time, the call of home ownership as a route to establishing independent wealth found renewed vigor in the United States and elsewhere, even as the middle class saw its gains hollowed out by wage stagnation and price inflation (particularly hard-to-prospect costs for things like college tuition and healthcare). There was enough borrowable money to go around. Everybody could get a loan.

The amount of investable money that came available each month was staggering. Attracted by all the opportunity, asset management exploded across a bewildering array of actors, from investment trusts to commercial banks, investment banks, pension funds, private banking organizations, insurance companies, mutual fund companies and unit trusts. With so many layers between them and what their money was doing out there scattered across the bond market, many people had little idea what exactly they were investing in, and their closest fund manager or financial advisor might not really have known either. In more than a footnote to complexity and its unforeseeable interdependencies, British to-be pensioners ended up locking up a lot of their savings in BP shares, which halved in value in the two months following the Deepwater Horizon disaster alone. It led to calls from the *Daily Mail*, a tabloid newspaper in Britain, for President Obama to stop bullying BP in 2010, saying that British retirees were going to foot the bill.[13]

And how did the asset managers make money? The issuer of a collaterized debt obligation, which might typically be an investment bank or some other entity working in their stead, earned a commission when it issued the bond, and then earned management fees during the remainder of its life. The ability to earn money from originating and securitizing loans, coupled with the absence of any residual liability (you didn't own an asset you could lose, after all, like a house), skewed the incentives of banks and financial managers in favor of loan volume rather than loan quality. More was better. And the Gaussian copula was the great enabler. Trade anything you could get your hands on, any debt you can find, run it through the copula and you'd see how it did or would possibly do. In a sense it was a replay of "let's-securitize-everything-we-can, even securities themselves" enthusiasm that swept Enron in the late 1990s (see Chapter 7). Securitizations exploded, with everything from lotto winnings to proceeds from tobacco lawsuits being turned into securities that could be sold to the investing public.[14]

A world of distant assets, future bets, and seemingly virtual debts could be traded, hedged, securitized – and generate money for those organizations that did the trading (or at least make their accounting figures look really good). Everybody used the Gaussian copula, from bond investors to Wall Street banks to ratings agencies and financial regulators. In complexity science this is known as positive feedback: more leads to more. Everybody started using the copula, because everybody started using it. Organizations that didn't use it ran the risk of being left behind in the skyrocketing securities trade. Once they embraced the copula as a basic instrument to assess risk, they no longer needed to look at the quality of any underlying securities. All they had to do was look at a correlation number, and out would come a rating telling them how safe or risky a tranche was.

In finance, it is impossible to ever reduce risk outright. We can only try to set up a market in which people who don't want risk sell it to those who do. But in the collaterized debt obligations market, people used the Gaussian copula model to convince themselves they didn't have any risk at all. As said, there was no single designer behind the creation and explosive growth of debt trading. The Gaussian copula wasn't the designer either. It was, if anything, a multiplier – something that made the trading easier, and attracted more trading-on-trading (that is, hedging other people's hedges, taking out loans to invest in other people's loans) something it could do by collapsing the complexity of the risk in these deals into a single, elegant number. It was easy to convince people that good things move together. If house prices go up, house prices go up. Bubbles grow when everybody is doing what everybody is doing. Again, that is positive feedback. But this also works the other way: bad things also tend to move together. Bubbles get punctured when everybody is doing what everybody is doing. If one mortgage defaults, then it is not unlikely that 100 others will too. As the saying goes: in a crisis, all correlations go to 1.[15]

When house prices stopped rising, less became less. Once again, there was positive feedback. Fewer people wanted to lend money because fewer people wanted to lend money. The huge web of interconnections and interdependencies that had grown on the back of the unfathomable amount of available bond market money now started backfiring, reverberating in the opposite direction. It was a feedback loop that made the world run dry of borrowable money in very short order. And it triggered an economic crisis and a huge recession.

After the fact ("ex post"), it all seems like a really bad idea. But that is the language of rational choice theory. After all, the badness of the idea became obvious only once we could see how it turned out. And how it turned out depended not on the Gaussian copula function per se. It depended on the world the function was released into, and the mass of relationships and interdependencies that started growing around it as people saw how it could work for them locally, relative to their goals, knowledge and focus of attention.

COMPLEXITY, LOCALITY AND RATIONALITY

David Li could hardly have known that his formula would ride a wave of enthusiasm and success that itself was ballooning on the back of a bond market swollen without precedent, and he could hardly have known that it would become one enabler of the failure of a complex system. According to cybernetics, an approach that is closely related to complexity theory and systems thinking, knowledge, like that of David Li or the asset manager selling an investment to somebody, is inherently subjective.[16] Direct, unmediated (or objective) knowledge of how a whole complex system works is impossible. Knowledge is merely an imperfect tool used by an intelligent agent to help it achieve its personal goals. An intelligent agent not only doesn't need an objective representation of reality in its head (or in its computer model) – it can't ever achieve one. The world is too large, too uncertain, too unknowable. For cybernetics, it can be argued that an intelligent agent doesn't have access to any "external reality." If there was an external, objective reality, the intelligent agent couldn't know it:

> The agent can merely sense its inputs, note its outputs (actions) and from the correlations between them induce certain rules or regularities that seem to hold within its environment. Different agents, experiencing different inputs and outputs, will in general induce different correlations, and therefore develop a different knowledge of the environment in which they live.[17]

The frame of reference for understanding people's ideas and people's decisions, then, should be their own local work context, the context in which they were embedded, and from whose (limited) point of view assessments and decisions were made. This is not the same as some globalize, after-the-fact ideal. A challenge is to understand why assessments and actions that from the outside look like really bad ideas appeared, from the inside, unremarkable, routine, normal, or systematically connected to features of the work environment we have put people in. In psychology, this has become known as the local rationality principle: people are doing what makes sense given the situational indications, operational pressures, and organizational norms existing at the time.

The local rationality principle grew out of a dissatisfaction with attempts to understand human functioning by reference to an ideal world in which people have access to all information and all the time they need to reach a good decision – the rational choice position. Rational choice here means that decision-making processes can be understood with reference to normative, optimal strategies. Strategies may be optimal when the decision-maker has perfect, exhaustive access to all relevant information, takes time enough to consider it all, and applies clearly defined goals and preferences to making the choice.

Rational choice theory sees bad decisions, then, as stemming from a decision-maker who deviates from this rational norm, this ideal. If the starting point for explaining human behavior is a rationalist norm, then any decision that deviates from that norm can be explained only on the basis of irrationality or limited awareness of the decision-maker. In other words, if people do not behave formally according to rational theory, it is because they either didn't get all the relevant information together or because they were irrational in their sorting and selecting of the best decision alternative. Either way, people can be reminded to be more rational, to try harder, to take more time to gather and consider all information, to be more motivated to do a reasonable job, to follow the established norms that exist in their environment. Perhaps more training or procedural guidance will help.

Rational decision-making requires a massive amount of cognitive resources and plenty of time. It also requires a world that is, in principle, completely describable. Complexity denies the possibility of all of these. In complex systems (which our world increasingly consists of) humans could not or should not even behave like perfectly rational decision-makers. In a simple world, decision-makers can have perfect and exhaustive access to information for their decisions, as well as clearly defined preferences and goals about what they want to achieve. But in complex worlds, perfect rationality (that is, full knowledge of all relevant information, possible outcomes, and relevant goals) is out of reach. There is not a single cognitive system in the world (either human or machine) that has sufficient computational capacity to deal with it all. From the 1970s onward, an increasing amount of psychological work pointed to the various ways in which human rationality is local, not global (and not "perfect" in that sense). Herbert Simon observed how the capacity of the human mind for formulating and solving complex problems is very limited compared to the potential size of those problems, and that, as a result, human rationality is bounded. Objective (or perfect, or global) rationality is cognitively impossible (in fact, it is impossible for a computer too: decision problems always have more aspects and dimensions than can be enumerated in written code).

In complex systems, decision-making calls for judgments under uncertainty, ambiguity and time pressure. In those settings, options that appear to work are better than perfect options that never get computed. Reasoning in complex systems is governed by people's local understanding, by their focus of attention, goals, and knowledge, rather than some (fundamentally unknowable) global ideal. People do not make decisions according to rational theory. What matters for them is that the decision (mostly) works in their situation. What matters is whether the situation still looks doable, that they are getting out of it what they want, and that their decisions are achieving their goals as far as they understand. Yet in a complex system, of course, that which can work really nicely locally, can make things fail globally. Optimizing decisions and conditions locally can lead to global brittleness.

COMPLEXITY AND DRIFT INTO FAILURE

In complex systems, decision-makers are locally rather than globally rational. But that doesn't mean that their decisions cannot lead to global, or system-wide events. In fact, that is one of the properties of complex systems: local actions can have global results. Because of the multitude of relationships, interconnections and interdependencies, they possess the ability to propagate the influence of a local decision through themselves. Of course, there will be modulation or amplification or deflection along the way, simply as a result of the journey through the complex system. But with so many interacting and interdependent agents, or components, what happens in one corner, or what gets decided in one corner, can ripple through the system beyond the possible predictions or knowledge of the original decision-maker. The Gaussian copula of David Li is one example.

This book focuses on the reverberations of local decisions through a complex system over time. One of the patterns that can become visible, particularly after the fact, is that of drift into failure. Local decisions that made sense at the time given the goals, knowledge and mindset of decision-makers, can cumulatively become a set of socially organized circumstances that make the system more likely to produce a harmful outcome. Locally sensible decisions about balancing safety and productivity – once made and successfully repeated – can eventually grow into unreflective, routine, taken-for-granted scripts that become part of the worldview that people all over the organization or system bring to their decision problems. Thus, the harmful outcome is not reducible to the acts or decisions by individuals in the system, but a routine by-product of the characteristics of the complex system itself.

Here is how that may work. Decisions make good local sense given the limited knowledge available to people in that part of the complex system. But invisible and unacknowledged reverberations of those decisions penetrate the complex system. They become categories of structure, thought and action used by others (for example, a Gaussian copula becomes *the* risk assessment tool), and they limit or amplify the knowledge available to others (for example, a single risk number). This helps point people's attention in some directions rather than others, it helps direct and constrain what other people in the complex system will see as sensible, rational or even possible. Wherever we turn in a complex system, there is limited access to information and solutions. Embedded as they are in a complex system, then, individual choices can be rendered unviable, precluding all but a few courses of action and constraining the consideration of other options. They can even become the means by which people discover their or their organization's very preferences.[18]

Adaptive responses to local knowledge and information throughout the complex system can become an adaptive, cumulative response of the entire system – a set of responses that can be seen as a slow but steady drift into failure. We can distinguish various features in this drift. A short preview is below, but they will be worked out in greater detail in the next chapter.

First, complex systems can exhibit tendencies to drift into failure because of uncertainty and competition in their environment. Adaptation to these environmental features is driven by a chronic need to balance resource scarcity and cost pressures with safety. Resources necessary to achieve organizational ends can be scarce because their nature limits supply, because the activities of regulators, competitors, and suppliers make them scarce, or because there might be pre-existing commitments that keep the organization from getting the resources it wants.[19] BP, for example, had fallen behind schedule in its Deepwater Horizon operations, and was quickly going over budget, paying 500,000 dollars per day to lease the rig from Transocean. By the day of the explosion, it was 43 days late in starting a new drilling job, a delay that had already cost the company more than 21 million dollars. Such production and schedule pressure, cost issues and resource scarcity translate into multiple, conflicting goals of a complex system that get resolved or reconciled or balanced in thousands of larger and little trade-offs by people throughout the system and its surrounding environment everyday.

This leads to the second feature: drift occurs in small steps. This can be seen as decrementalism, where continuous organizational and operational adaptation around goal conflicts and uncertainty produces small, step-wise normalizations of what was previously judged as deviant or seen as violating some safety constraint. Each decrement is only a small deviation from the previously accepted norm, and continued operational success is taken as a guarantee that the small adaptations are safe, and will not harm future safety. Decisions set a precedent, particularly if they generate successful outcomes with less resource expenditure. People come to expect and accept deviance from what may have been a previous norm. They have come to believe that things will "probably be fine" and that it will do "a good job":

> In one instance, four days before the April 20 explosion, one of BP's operations drilling engineers sent an email to a colleague noting that engineers had not taken all the usual steps to center the steel pipe in the drill hole, a standard procedure designed to ensure that the pipe would be properly cemented in place. 'Who cares, it's done, end of story, we'll probably be fine and we'll get a good cement job,' he wrote.[20]

Since its operations on the Alaska North Shore (itself an environment of intense resource scarcity because of geographical isolation, despite a lack of immediate neighbor competition), BP had gradually come to accept a maintenance practice of "run to failure" where a part would not be replaced preemptively. Arguing for preventive maintenance would increasingly have fallen outside the range of choices that decision-makers could explain as rational. The use of language such as "run to failure" and its repetitive reproduction throughout the organization, legitimized a build-up of what others would later see as unacceptable maintenance backlogs. Remember the audit report that said that BP had "neglected" to clean and check pressure valves, emergency shutoff valves, automatic emergency shutdown mechanisms and gas and fire safety detection devices? Inspection reports had been "pencil-whipped" or approved fraudulently without the part having been seen or meeting necessary criteria. For those on the inside, however, the use of language and repeatedly successful decisions would have meant that there was no backlog, and no maintenance need, and no fraudulent approval – after all, the concerned parts hadn't run to failure yet.

In a complex system, however, doing the same thing twice will not predictably or necessarily lead to the same results. Past success cannot be taken as a guarantee of future success or safety. It cannot be taken as a source of confidence that the adaptive strategies to balance resource scarcity, competition and safety will keep on working equally successfully in all future conditions. The practice of parts to failure may not have resulted in large safety problems yet, but that doesn't mean it never will. Of course the belief might be that running individual parts to failure is merely a reliability problem, not a safety problem – if the part is seen in isolation. But parts do not work in isolation. Parts work in the context of other parts and systems and process expectations: their failure (a problem of reliability) can reverberate throughout the web of relationships that this part has with other parts and systems, and cascade into system failures. *That* is a safety problem.

Third, complex systems are sensitively dependent on initial conditions. A very small decision somewhere in the beginning of the life of the complex system (for example, putting the γ at the end of the Gaussian copula) can blow into huge effects later on. Had David Li decided to replace the "equal" sign with something that would indicate approximation rather than equality, the Gaussian copula may not have provided the basis for as much confidence in its final number as it did. This confidence, and the ease of the production of the number, after all, helped accelerate its use way beyond Li's wildest expectations. In other words, the potential for drift into failure can be baked into a very small decision or event. This may have happened somewhere way back, way at the beginning, even before the system was as complex as it may now be.

Take BP again. Even while struggling to stop the oil spill in the Gulf of Mexico, the company was pursuing projects in other parts just off the shore of the U.S. In the Beaufort Sea north of Alaska, for example, it was getting ready to drill two miles down and then six miles horizontally to reach a putative 100-million barrel oil deposit under U.S. waters. Offshore drilling, however, had been hit by a government moratorium in response to the Gulf spill. Yet BP's project, called "Liberty," was not included in the moratorium. The reason was that it was not formally classified as "offshore." Rather than putting in a traditional steel-legged or floating rig, BP had decided to construct an artificial island some three miles off the coast in the Beaufort Sea by heaping gravel into a pile in 22 feet or 7 meters of water. This had a technical rationale: a traditional rig would have difficulty withstanding the pressure of ice floes butting up against it.

But classifying this drill site as "onshore" could be one of those initial conditions that create sensitive dependency. On the one hand the classification had all kinds of consequences for subsequent regulatory and environmental approval processes (generally making things easier). On the other hand, the "island" would be way too tiny and porous to contain any oil spill. Oil from a spill would immediately flow into the surrounding sea (used by seals, polar bears, whales and everything down the food chain from there on): an on-shore project with immediate off-shore consequences, in other words, or sensitive dependency on initial conditions. Even though BP's Gulf disaster was the reason behind the offshore oil drilling moratorium, it was left as the only company in the Beaufort Sea to drill offshore, because its operation had been classified as onshore.

Fourth, complex systems that can drift into failure are characterized by unruly technology. Unruly technology introduces and sustains uncertainties about how and when things may develop and fail. These are uncertainties that cannot typically be reduced with Newtonian logics of linear calculation (for example, fault trees). The Gaussian copula was one such piece of unruly technology. It was impossible to say with certainty how it would operate, and even David Li was leery of those who relied on its results too much. Even the very inputs to the Gaussian copula were rife with uncertainty, something that was not reduced but rather obscured by crunching them through the terms of the formula. Deepwater drilling relies on hugely unruly technology. The rigs are monstrously large, and their size alone tends to generate complexity. In the search for end-stage oil, drilling in shallow waters has begun to be gradually replaced by deep-sea drilling where the rig may reach a mile below the surface, and then probe another three miles below the ocean floor to get to oil. In the Liberty project, technology has been acknowledged to be even more experimental and potentially unruly with its two vertical and six horizontal miles of pipe. Even BP itself has promoted the project as pushing the boundaries of drilling technology. Drilling down and then sideways under an ocean floor can prevent rigs from having to be placed in impossible or sensitive areas. But it makes the technology more prone to the sort of gas kicks that can turn into blowouts that lay behind the demise of the Deepwater Horizon. So much force is needed to drill over such long distances that BP invested in a new steel alloy for the drill pipe. The technology, called "extended reach drilling," is known to push the envelope.[21]

And finally, complex systems, because of the constant transaction with their environment (which is essential for their adaptation and survival) draw in the protective structure that is supposed to prevent them from failing. This is often the regulator, or the risk assessor, the rating agency. The protective structures (even those inside an organization itself) that are set up and maintained to ensure safety, are subject to its interactions and interdependencies with the operation it is supposed to control and protect. Imperfect knowledge, lack of information or access, and decisions that make local sense, affect the relationship with the regulator as much as they affect the complex functioning of the operator itself. Because of this, the protective structure can actively contribute to drift into failure, rather than just not intervening when it should. For example, a regulator can make certain risk assessments necessary by legally requiring them or unnecessary by waiving them. In either case, results or their absence are legitimized. Rating agencies did this, the aviation regulator of the next chapter did this, and the internal risk management department of Enron did this.

Deepwater drilling, which took both the technology and the regulatory apparatus of shallow water drilling into much lesser-known territory, was governed by ad-hoc exemptions as much as by rules. The process was called "alternative compliance" by which deep-water rig technology was approved piecemeal and regulators cooperated with industry groups to make small adjustments to decades-old rules and guidelines. Of the roughly 3,500 rigs in the Gulf, fewer than 50 operate in waters deeper than 1,000 feet, and the deeper the water, the less certainty there was about the applicability of the rules and guidelines. In the case of deepwater drilling by BP, the organization was exempted in 2008 from an environmental review that would have had them filing a plan on how they would clean up a major oil spill. Also, they did not need a fire department. Testing a blowout preventer (in hindsight a crucial piece of technology) was approved at a lower pressure than was federally required. In the BP Liberty project in the Alaskan Beaufort Sea, regulators allowed BP in 2007 to write its own environmental review, as well as its own consultation documents relating to the Endangered Species Act. The environmental assessment was taken away from the Mineral Management Service unit that normally handles such reviews, probably because its result would have been too negative. The final 2007 U.S. government environmental review of BP's own assessment was virtually identical to that produced by BP itself: its text was in large part a matter of copy-and-paste.

The protective structure also actively contributes to complexity because it often consists of a number of players, some of whom may have conflicting goals. One of the main regulators, the Minerals Management Service, or MMS, had an explicit mandate to encourage the production of oil, while also having to collect fines from drilling companies if they violated the law. The agency is the second-largest source of federal income after the Internal Revenue Service. In 2009 it received 264 million dollars from offshore oil and gas production, mainly from royalties and from leasing blocks of sea floor. No one agency was solely responsible for guaranteeing overall rig safety. Self-regulation or voluntary oversight, as with Wall Street during the collaterized debt obligation hausse, took over from externally dictated regulation and close government monitoring. Oil rig inspections by the government in the Gulf took the form of helicopter visits to drilling platforms. With only 60 inspectors to oversee 4,000 rigs, regulators could do little more than sift through documentation and sit through presentations during their site visits. The complexity and the expertise required to make sense of the operations would have defeated attempts at meaningful regulation in any case. Such practices, assumptions and expectations and their legitimized, legal results got taken as a basis for confidence that risk was under control, that organizations were striking a good balance between safety and production.

A GREAT TITLE, A LOUSY METAPHOR

Drift can be seen as an inexorable slide towards ever smaller margins, toward less adaptive capacity, towards a growing brittleness in the face of safety challenges that the world may throw at the organization. What is hidden behind that image, however, is that drift itself is the result of constant adaptation and the growing or continued success of such adaptation in a changing world. It is the result of fine-tuning the system's computational rules and constraints (which in complex systems, as we will see, is thought to be optimized at the edge of chaos). Drift, in other words, is not just a decrease of a system's adaptive capacity. It is as much an emergent property of a system's adaptive capacity. Congress diverting funds to the International Space Station meant a changing environment for Space Shuttle Operations (which created the side effect of increasing pressure on launch schedules so as to get building materials up to the Space Station). This challenge in a changing environment was met by the system's adaptive capacity: it got absorbed, adapted around, in part by further fine-tuning or finessing of various rules and constraints that, eventually, influenced turn-around times between Shuttle launches.

It takes more than a few sudden decisions by badly informed or badly moralized or motivated people to push a system over the edge into failure. Failure in complex systems takes more than that, and it takes longer than that. The first family of explanations introduced in this chapter suggests that it is in individual actions and broken components that we should look for the causes of disaster and harm. This first family of explanations, of Newton and Descartes, is over-used, over-applied, and over-developed. But it is far from well-suited for explaining success and failure in most systems in our world today. The second family, of complexity, is under-used, under-applied and under-developed. Which is why this book is first about the contrast, and then about the second family. What are its ideas? What is the theory behind it? What new questions does it allow us to ask?

It is precisely because a system is a complex web of relationships that it can be creative and adapt to environmental challenges. This is how it survives – which is of course the story of evolution. Such survival should not be mistaken for progress (or regress). Complexity doesn't allow us to think in linear, unidirectional terms along which progress or regress could be plotted. Drift into failure surely invites the idea of regress (less margin, less room to recover, a decline in norms, a regression from higher forms of organizational stature to lower ones) as its central doctrine, and encourages the interpretation of history in sequences (as in the increase in lubrication intervals in Alaska 261, see Chapter 2). But complexity suggests that it is about constant adaptation to what happens *now* (or shortly ago), and any larger historical story that we put on it is ours, not the system's.

WHY WE MUST NOT TURN DRIFT INTO THE NEXT FOLK MODEL

Drift into failure may be an evocative title for a book – it calls to mind something inevitable, something slow and unstoppable. It may suggest that failure is always an option. And in complex systems, failure *is* always an option. But it is a better title than a metaphor. After all, adaptive behavior in complex systems probably moves in all kinds of ways, not just in one direction. Failure is one option; other outcomes are always possible too. A particular direction becomes visible only from the position of retrospective outsider, when we are able to lift ourselves out and look back on what has happened. Thus it may only be possible in hindsight to see how decisions and trade-offs accumulated and reverberated to create a particular trajectory toward an accident. The drift is thus a construct in retrospect, after the failure has already happened. Or, at the most, it is a construct that can be applied to a complex system from the outside. From the inside, drift is invisible. That actually also offers a lever for thinking about preventing drift and preventing it from becoming failure (see Chapter 7). Outside perspectives and new perspectives are important, as is the diversity of perspectives and taking a long view of things.

There is a deeper reason to be suspicious of drift into failure as a metaphor for what happens with and inside organizations. This has to do with its sense of inexorable decline or regress into something worse than what was before. This would be the inverse of the narrative that the West has always liked to tell about itself: the story of unstoppable progress. Newton and Descartes lie at the roots of this narrative (and, as Chapter 3 will show, society had good reason to put them there). Progress as a central concept to organize Western history around reached its peak in the nineteenth century with the triumphs of the Industrial Revolution and Victorian imperialism and expansionism. It was made literal in a linear image in the U.S.A. where white settlers pushed the frontier ever further west. The further, the more progress. The fuller these settlements became, and the more supported by Industrial Revolution technology (for example, steam trains), the more progress people saw.

More recent decades show, at least in places, a reversal of this central Western narrative. Progress as the prevailing doctrine in the interpretation of historical sequence is being replaced, at least in some places, by a doctrine of regress. This has been infused by pessimism about human nature after having come out of the most murderous of centuries (the twentieth). People see regression in steady ecological destruction, in the decline of America and its market-driven capitalist model, in secularization and in the depreciation of the West as a whole. Specifically, we find seductive the idea of moral fragmentation, of growing societal anomie, a lack of social or ethical standards in individuals and groups.[22] This, in many societies, is seen as drift into failure on a grand scale: the decline of norms, the irrelevance or unenforceability of compliance, the collapse of traditional systems of normative observance. In other words, a regressive metaphor for thinking about safety (like drift into failure) could tumble into well-prepared psychological soil.

But such recognition should not be confused with the truth value of the metaphor. Evolution has no direction, at least no visible direction when you're in the middle of it. Evolution, progress and regress in complex systems are perhaps nothing more than an illusion, a cultural, psychological imposition that comes from us, not from the system or its history. Stephen Jay Gould called progress "a noxious, culturally embedded, untestable, nonoperational idea that must be replaced if we wish to understand the patterns of history."[23] If history is seen as progressive (or, for that matter, regressive, as in drift), then it becomes all too easy to see it as directed, as following a vector – going at a particular speed in a particular direction. One consequence is that it becomes tempting, once again, to see a hand behind that vector. A "cause" for its speed and its direction. Which would be a Newtonian idea, an idea that clashes with complexity.

The risk of a folk model exists when we grasp a label only superficially and subsequently start to recognize it everywhere in our world. This book wants to avoid introducing a folk model. So it is not an attempt at exegesis. It is not claiming that we can read drift and normal decisions and their resulting drift (or, on the other hand, fraudulence and amorality and their bad decisions) *out of* the story we encounter. Exegesis makes the assumption that the essence of a story is already in the story. It is there, waiting and ready-formed for us to discover. All we need to do is read the story well, apply the right method, use the correct analytic tools, set things in context, and the essence will be revealed to us. Such structuralism sees the truth as being "behind" or "within" a text. Some stories will show a trajectory of drift, others will not. Simply because the drift is already there or it is not – all we need to do is look whether we can find it.

Rather, in encouraging us to think about drift into failure, this book makes an appeal to *eisegesis*, of us reading something *into* the story. Any directions or vectors that we discern in such adaptation are of our making – Gould's "culturally embedded idea" of progress (or regress). Foucault too was always skeptical of linear trajectories (either up or down) as descriptive of the social order, including the process of, and perspectives on, knowledge or science. His position, and in many ways the position of complexity, is that of post-structuralism, the broad intellectual movement that swept social and other sciences synchronous with the growth of complexity theory and systems thinking in the latter half of the twentieth century. Post-structuralism stresses the relationship between the reader and the text as the engine of "truth." Reading in post-structuralism is not seen as the passive consumption of what is already there, provided by somebody who possessed the truth and is only passing it on. Rather, reading is a creative act, a constitutive act, in which readers generate meanings out of their own experience and history with the text and with what it points to.

As post-structuralism sees it, author and reader aren't very different at all. Even authors write within a context of other texts (data from a cockpit voice recorder, for example, or documents about an organization's quality assurance system) in which choices are made about what to look at, what not – choices that are governed by the author's own background and institutional arrangements and expectations. The author is no longer a voice of truth. Text, or any other language available about events or accidents, has thereby lost its stability. Nietzsche, for example, would have been deeply mistrustful of the suggestion that while everybody has different interpretations, it is still the same text. He, and post-structuralism in general, does not believe that it is a single world that we are all interpreting differently, and that we could in principle reach agreement when we put all the different pictures together. More perspectives don't mean a greater representation of some underlying truth.

In a complex system the contrary seems to be at work instead. More perspectives typically mean more contradictions. Of course, there might be some partial overlap, but different perspectives on an event will create different stories that are going to be contradictory – guaranteed. The reason, says complexity science, is that complex systems can never be wholly understood or exhaustively described. If they could, they would either not be complex, or the entity understanding them would have to be as complex as the whole system. The complexity of the whole system would have to present *in that part*. Which the local rationality principle says is impossible. We (people, computer models, regulatory agencies, risk assessments) can only understand parts of a complex system, and pretty local parts at that. For what happens further away we have to rely on the perspectives and interpretations of others. And we can only learn about those interpretations from those people – or from yet other people. By the time we learn about them, such interpretations will likely have gotten modulated, adapted, suppressed, enhanced, changed.

Implicit in this stance is that there are no final conclusions. The events we want to study (just like any complex system) can never be fixed, tied down, circumscribed conclusively with a strong perimeter that says what is part of it and what is not. Even if the system has pretty much died, and all we have is archeological data (for example, after a plane crash which caused the airline to go defunct), it will never be possible to say that we have all the data, that we have every perspective, that we have considered all angles. We simply can't know.

We may never be able to successfully argue that some accidents are the result of drift into failure, and that others are not. Of course it would be convenient, and quite "scientific," perhaps, to carve up the world into categories and say, "this is a drift into failure accident, and this is not." But post-structuralism doesn't really allow us to say whether something *is* or *is not* in the text (in the accident, in the event) itself. It is up to the reader, the observer, to interact with the text – the event, the accident – to interrogate it, to read a particular background and questions into it, and then bring out what is relevant, what is "true" to that reader. By reading drift into a particular failure, we will probably learn some interesting things. But we will surely miss or misconstrue other things.

If we don't believe that a particular accident is the result of a drift into failure, but rather the result of a momentary, sudden stochastic contraction of elements in time and space that had no antecedent, no history, no diachronic trajectory, no path dependency – fine. We can certainly go ahead and build a persuasive story around that conviction. But by not reading drift into it, we might also miss things and misconstrue others. In the end though, whether there is drift into failure depends not on what is *in the story*, but on what is *in us*. It depends on what *we* bring to the story, what we read into it. It depends on how far we read, how deeply, how far back, what else we read. And it depends on the assumptions and expectations we bring about knowledge, cause, and ultimately morality.

REFERENCES

1   Rasmussen, J., and Svedung, I. (2000). *Proactive risk management in a dynamic society*. Karlstad, Sweden: Swedish Rescue Services Agency, p. 14.

2   Woolfson, C. and Beck, M. (eds.). (2004). *Corporate responsibility failures in the oil industry*. Amityville, NY: Baywood Publishing Company, Inc.

3   Epistemology is the branch of philosophy that asks how we know what we know. By epistemological space, I mean a space in which certain ideas are seen as logical or possible, while other ideas are seen as ridiculous or even fall completely out of consideration because the epistemological space contains no language, no knowledge to put them in.

4   This is of course unfair to both men, not only because their own ideas are more complex and conflicted and paradoxical than what I make them out to be, but also because subsequent people (e.g. Pierre Simon Laplace, 1749–1827) played a greater role in popularizing some of the ideas more then they themselves did. And, of course, there were people before them (like Plato, 429–347 BCE and Socrates, 469–399 BCE) who may have been even more instrumental in laying a philosophical

foundation for these ideas and their opposite. There is, however, a case to be made for the existence of a set of ideas that can roughly be called Newtonian–Cartesian, which is what I do in this book.

5   Page, S.E. (2008). Uncertainty, difficulty and complexity. *Journal of Theoretical Politics*, 20(2), 115–49, p. 115.

6   Broder, J.M., and Calmes, J. (2010, June 18). Chief of BP, contrite, gets a scolding by Congress. *International Herald Tribune*, p. 1.

7   Mufson, S., and Kumblut, A.E. (2010, June 15). Amid claims of BP's 'shortcuts,' Obama speech to stress action, prevention. *Washington Post*.

8   Mufson, S., and Kumblut, A.E. (2010, June 15). Ibid.

9   Mufson, S., and Kumblut, A.E. (2010, June 15). Ibid.

10 Green, J. (2003). The ultimate challenge for risk technologies: Controlling the accidental. In J. Summerton and B. Berner (eds.), *Constructing risk and safety in technological practice*. London: Routledge.

11 This is Paul Cilliers' pithy insight. See: Cilliers, P. (1998). *Complexity and postmodernism: Understanding complex systems*. London: Routledge.

12 Associated Press (2010, July 18). *Number of homeless families grew in 2009, report says*. International Herald Tribune, p. 4.

13 Broder, J.M., and Calmes, J. (2010, June 18). Ibid., p. 3.

14 McLean, B. and Elkind, P. (2004). *The smartest guys in the room: The amazing rise and scandalous fall of Enron*. New York: Penguin, p. 133.

15 As Snook explains in his 2000 book, sudden contractions or stochastic fits in system coupling (indeed the crises where all correlations go to 1) occur in systems that have undergone practical drift, where locally grown logics of action have made individual units slide apart. Once crisis, or a sudden contraction in coupling, puts them back together with demands for quick and safe cooperation, the fissures and gaps that have developed become acutely active and sometimes even visible, leading to misunderstandings, an increase in risk and to failure. See: Snook, S.A. (2000). *Friendly fire: The accidental shootdown of US Black Hawks over Northern Iraq*. Princeton, NJ: Princeton University Press.

16 Heylighen, F., Cilliers, P., and Gershenson, C. (1995). *Complexity and philosophy*. Brussels: Vrije Universiteit.

17 Heylighen, F., Cilliers, P., and Gershenson, C. (1995). Ibid., p. 8.

18 Vaughan, D. (1996). *The Challenger launch decision: Risky technology, culture and deviance at NASA*. Chicago: University of Chicago Press, p. 37.

19 Vaughan, D. (1999). The dark side of organizations: Mistake, misconduct, and disaster. *Annual Review of Sociology*, 25, 271–305.

20 Mufson, S. and Kumblut, A.E. (2010, June 15). Ibid.

21 Urbina, I. (2010, June 25). Some call a new BP project risky: 'On-shore' rig is located on an artificial island, and oversight is criticized. *International Herald Tribune*, p. 2.

22 Giddens, A. (1991). *Modernity and self-identity: Self and society in the Late Modern age*. London: Polity Press.

23 Gould, S.J. (1987). Opening remarks of the conference on evolutionary progress at Chicago's Field Museum. Quoted in Lewin, R. (1999). *Complexity: Life at the edge of chaos*, Second Edition. Chicago, IL: University of Chicago Press. p. 139.

2
FEATURES OF DRIFT

In the early afternoon of January 31, 2000, Alaska Airlines flight 261, a McDonnell Douglas MD-80 took off from Puerto Vallarta in Mexico, bound for Seattle. The pilots had just taken over the airplane from the incoming crew, who had nothing special to report about the status of the airplane.[1]

A bit into the flight, the pilots contacted the airline's dispatch and maintenance control facilities in Seattle on the radio. This was a shared company radio frequency between Alaska Airlines' dispatch and maintenance facilities at Seattle and its operations and maintenance facilities at Los Angeles International Airport (LAX). They had run into a pretty serious problem: the horizontal stabilizer, which helps control the aircraft's nose attitude while in flight, appeared to be jammed.

"Understand you're requesting diversion to LA?" Seattle maintenance asked the pilots at 3.50 p.m. "Is there a specific reason you prefer LA over San Francisco?"

"Well a lot of times it's windy and rainy and wet in San Francisco and uh, it seemed to me that a dry runway, where the wind is usually right down the runway seemed a little more reasonable."

A few minutes later, a dispatcher from Seattle provided the flight crew with the current San Francisco weather. The wind was light, out of the south (180 degrees), and the visibility was good (9 miles). But, the dispatcher added, "If you want to land at LA of course for safety reasons we will do that, we'll tell you though that if we land in LA, we'll be looking at probably an hour to an hour and a half we have a major flow program going right now," referring to air traffic control restrictions that would make it hard to get the aircraft out of Los Angeles again.

"I really didn't want to hear about the flow being the reason you're calling us, because I'm concerned about over flying suitable airports," the captain replied. He did, however, discuss with his first officer potential landing runways at SFO, and finding a discrepancy: "One eight zero at six ... so that's runway one six what we need is runway one nine, and they're not landing runway one nine."

"I don't think so," the first officer replied.

The captain then asked Seattle dispatch if they could "get some support" or "any ideas" from an instructor to troubleshoot the problem.

He received no response.

"It just blows me away," the captain then said to his first officer, "they think we're gonna land, they're gonna fix it, now they're worried about the flow. I'm sorry, this airplane's not gonna go anywhere for a while. So you know."

"So they're trying to put the pressure on you," a flight attendant replied.

"Well, no. Yeah."

The Seattle dispatcher had not given up on San Francisco. He informed the flight crew a few minutes later that the landing runways in use at SFO were 28R and 28L and that "it hasn't rained there in hours so I'm looking at ... probably a dry runway."

The captain replied that he was waiting for a requested center of gravity (CG) update (for landing), and then he requested information on wind conditions at Los Angeles. The dispatcher replied that the wind at LAX was out of the west (260°) at 9 knots.

Nine seconds later, the captain, comparing SFO and LAX wind conditions, told the SEA dispatcher, "versus a direct crosswind which is effectively no change in groundspeed ... I got to tell you, when I look at it from a safety point I think that something that lowers my groundspeed makes sense."

"That'll mean LAX then for you," the dispatcher replied. He then asked the captain to provide LAX operations with the information needed to re-compute the airplane's CG because "they can probably whip out that CG for you real quick."

"We're going to LAX," the captain then said. "We're gonna stay up here and burn a little more gas, get all our ducks in a row, and then we'll be talking to LAX when we start down to go in there."

It was now almost four in the afternoon.

Turning to LAX operations, the captain asked if they could "compute [the airplane's] current CG based on the information we had at takeoff," and asked them once again for the latest weather information at San Francisco. The weather was basically unchanged.

"That's what I needed. We are coming in to see you," the captain then told LAX operations. The first officer began giving LAX operations the information it needed to re-compute the airplane's CG for landing.

Then, at 4.07 p.m., a mechanic at Alaska Airlines' LAX maintenance facility contacted the flight crew on the company radio frequency and asked, "Are you [the] guys with the uh, horizontal stabilizer situation?"

"Affirmative," the captain replied.

The mechanic, referring to the stabilizer's primary trim system, asked, "Did you try the suitcase handles and the pickle switches?"

"Yeah. We tried everything together," the captain said, adding: "We've run just about everything. If you've got any hidden circuit breakers we'd love to know about them."

"OK, I'll look at the circuit breaker guide," the mechanic replied, "just as a double check." He then asked the flight crew about the status of the alternate trim system.

"It appears to be jammed … the whole thing. The AC load meter spikes out when we use the primary, we get AC [electrical] load that tells me the motor's trying to run but the brake won't move it. When we use the alternate, nothing happens."

"You say you get a spike … on the meter up there in the cockpit when you try to move it with the primary, right?"

In the meantime, the captain had turned to the first officer in the cockpit. "I'm going to click it off," he said, talking about the autopilot. "You got it."

"Okay," the first officer said.

Then the captain reiterated to the LAX mechanic that the spike occurred "when we do the primary trim but there's no appreciable change in the electrical when we do the alternate."

The mechanic, not able to figure things out from these reports from a distance, replied that he would see them when they arrived at the LAX maintenance facility.

"Lets do that," the captain said.

The autopilot was still on at this time, and the captain said to the first officer: "This will click it off." According to the Flight Data Recorder (FDR) data, the autopilot was then disengaged at 1609:16.

At the same time, the cockpit voice recorder (CVR) recorded the sound of a clunk, followed by two faint thumps in short succession at 1609:16.9. The CVR recorded a sound similar to the horizontal stabilizer-in-motion tone.

"You got it?" the captain asked.

The autopilot was disengaged again and the airplane began to pitch nose down, starting a dive that lasted about 80 seconds as the airplane fell from about 31,000 to between 23,000 and 24,000 feet.

"It got worse," the captain managed to say. "You're stalled."

Airframe vibration was getting louder now.

"No, no, you've got to release it, you've got to release it."

A click sounded.

"Help me back, help me back."

"Okay," the first officer responded.

"Center, Alaska two sixty one we are, ah, in a dive here," the captain told the air traffic controller. "And I've lost control, vertical pitch."

The overspeed warning came on, and continued for the next 33 seconds.

"Alaska 261, say again?" the controller asked in disbelief.

"Yeah, we're out of twenty-six thousand feet, we are in a vertical dive … not a dive yet … but uh we've lost vertical control of our airplane." Not much later, the captain added, "Just help me."

For a moment, things seemed to get better. "We're at twenty-three seven, request, uh … " And then the captain added, "Yeah, we got it back under control here."

The first officer didn't think so. "No we don't," he immediately told the controller. Turning to the captain, he suggested: "Let's take the speedbrakes off."

"No, no leave them there. It seems to be helping." The speedbrakes, which come out of the top of the wings to help get rid of airspeed, were still out as a response to the overspeed from the first dive.

"Okay, it really wants to pitch down," the captain said. Transmitting to the controller, the captain said that they were at "Twenty-four thousand feet, kind of stabilized."

Three seconds later, he added, "We're slowin' here, and uh, we're gonna uh do a little troubleshooting, can you give me a block [altitude] between uh, twenty and twenty-five?"

It was 4.11 p.m ... The airplane's airspeed had decreased to 262 knots, and the airplane was maintaining an altitude of approximately 24,400 feet with a pitch angle of 4.4°. The controller assigned flight 261 a block altitude of between FL 200 and 250. Between about 130 and 140 pounds of pulling force was required to recover from the dive and keep the aircraft level.

"Whatever we did is no good," the first officer said. "Don't do that again."

"Yeah, no, it went down. It went to full nose down."

"It's a lot worse than it was?"

"Yeah, we're in much worse shape now. I think it's at the stop, full stop. And I'm thinking, can it go any worse? But it probably can, but when we slowed down ... Lets slow it, lets get down to two hundred knots and see what happens."

The captain turned to LAX maintenance again. "We did both the pickle switch and the suitcase handles and it ran away full nose trim down. And now we're in a pinch. So we're holding, we're worse than we were." The captain explained that he was reluctant to try troubleshooting the trim system again because the trim might "go in the other direction."

"Okay, well, your discretion. If you want to try it, that's okay with me if not that's fine. Um, we'll see you at the gate."

"I went tab down ... right?" the captain explained to the mechanic, "and it should have come back instead it went the other way."

"You want to try it or not?" the captain then asked the first officer.

"Uh no. Boy, I don't know."

About 120 pounds of pulling force was being applied to the pilots' control columns at this point, just to keep the airplane level.

At a quarter past four, the Los Angeles controller instructed the flight crew to contact another controller on a different radio frequency, which the flight crew acknowledged.

"We're with you, we're at twenty-two five," the first officer radioed to the new controller, "we have a jammed stabilizer and we're maintaining altitude with difficulty. Uh, but, uh we can maintain altitude we think ... Our intention is to land at Los Angeles."

The controller cleared the airplane direct to LAX and asked, "You want a lower altitude now, or what do you want to do sir?"

The captain replied this time. "I need to get down about ten , change my configuration, make sure I can control the jet and I'd like to do that out here over the bay if I may."

The Los Angeles ARTCC controller issued flight 261 a heading of 280° and cleared the flight to descend to 17,000 feet.

"Two eight zero and one seven, seventeen thousand, Alaska two sixty-one," the captain said. "And we generally need a block altitude."

"Okay," the controller said, "and just, um, I tell you what – do that for now sir, and contact LA center on one three five point five. They'll have further instructions for you sir."

"Okay," the first officer said to the new controller, "thirty-five five, say the altimeter setting."

"The LA altimeter is three zero one eight."

"Thank you," the first officer said.

This would be the last radio call made from flight 261.

After the radio transmission, the captain told a flight attendant that he needed "everything picked up" and "everybody strapped down." He told the first officer that he was going to stop pulling so hard on the control column: "I'm going to unload the airplane and see if we can ... we can regain control of it that way."

The flight attendant came back with bad news. "Okay, we had like a big bang back there."

"Yeah," the captain said. "I heard."

He turned to his first officer again. "I think the stab trim thing is broke."

He then repeated to the flight attendant to make sure the passengers were "strapped in now," adding "because I'm going to release the back pressure and see if I can get it ... back."

"Give me slats extend," he asked the first officer.

It was eighteen minutes past four.

"I'm test flying now," the captain said.

More flaps were ordered, down to 11 degrees.

"Its pretty stable right here ... see? But we got to get down to a hundred and eighty knots."

Not much later, he asked the first officer, "Okay, bring the flaps and slats back up for me," The first officer complied.

"What I want to do ... is get the nose up ... and then let the nose fall through and see if we can stab it when it's unloaded."

"You mean, use this again?" the first officer asked. "I don't think we should ... if it can fly."

"It's on the stop now, it's on the stop," the captain answered.

"Well, not according to that, it's not," the first officer said, adding, "the trim might be, and then it might be uh, if something's popped back there ... it might be mechanical damage too. I think if it's controllable, we ought to just try to land it."

"You think so? Okay, lets head for LA."

About 5 seconds later, the CVR recorded four distinct thumps.

"You feel that?" the first officer asked.

"Yeah. Okay, give me slats."

Slats were extended again.

Then an extremely loud noise sounded, background noise started increasing and loose articles started moving all through the cockpit.

The airplane pitched down violently, at a rate of nearly 25° per second. The airplane started rolling left wing down, and the rudder was deflected back to the right.

The nose was now pointing almost straight down, at 70° and the aircraft kept rolling.

"Mayday," the first officer said, but did not manage to push the transmit button.

"Push and roll, push and roll," the captain said.

The dive diminished, but the airplane was now upside down, and had descended to 16,420 feet. The airspeed had decreased to 208 knots.

"Okay," the captain said, "we are inverted ... and now we got to get it."

Using rudder and ailerons, he tried to get the airplane upright again.

"Push, push, push ... push the blue side up," the captain counseled. "Okay, now lets kick rudder ... left rudder, left rudder."

"I can't reach it," the first officer said.

"Okay, right rudder ... right rudder. Got to get it over again." Then, acknowledging that they were now not diving as badly, he added, "at least upside down we're flying."

But not for much longer. With the airflow to them considerably disturbed, both engines now stalled and spooled down.

"Speedbrakes," the captain said.

"Got it," the first officer answered.

There was no more space for any recovery, and the aircraft would not have allowed it in any case.

"Ah, here we go," the captain concluded.

Twenty-one minutes past four, the airplane hit the Pacific Ocean near Port Hueneme, California. It came apart on impact and most of it disappeared below the surface. Pieces of the airplane wreckage were found floating on and beneath the surface of the ocean. The two pilots, three cabin crewmembers, and 83 passengers on board were all killed, and the airplane was destroyed by impact forces.[2]

The pilots had been confronted with one of the scariest experiences possible: an airplane that is no longer controllable. Whatever they did with their controls, it was to no avail. After a period of uncertainty and trouble-shooting, the airplane had suddenly plunged down, straight toward the ocean.

In the two weeks following the accident, remote-operated vehicles and sonar equipment was used to search the debris and load up and tow the large pieces of wreckage to the surface. A commercial trawler was used to drag the ocean bottom and gather the final smaller pieces. Eighty-five percent was recovered. It was all brought to Port Hueneme.

THE BROKEN PART

As the investigation examined airplane parts from the sea floor and matched the wreckage with data traces from the cockpit voice recorder and the flight data recorder, the culprit became obvious: the jackscrew-nut assembly that holds the horizontal stabilizer had failed, rendering the aircraft uncontrollable. The broken part had been found.

Like (almost) all aircraft, the MD-80 has a horizontal stabilizer (or tailplane, or small wing) at the back that helps direct the lift created by wings. It is this little tailplane that controls the aircraft's nose attitude: without it, controlled flight is not possible. The tailplane of the MD-80 controls nose attitude in two ways. At the back end of it, there's a control surface (the elevator) that connects directly to the pilots' control yokes in the cockpit. Pull the yoke in the cockpit back, the elevator angles up and the airplane's nose moves up in return.

The story of Alaska 261, however, is about another part of the horizontal stabilizer. The whole stabilizer itself can angle up or down in order to trim the nose up or down. As fuel is used up during the flight, or wing flaps are extended and retracted, or as a catering trolley moves up and down the aisle inside the aircraft during flight, the flight characteristics of the airplane change (for example, its center of gravity or the center of its lifting force shifts). To prevent the airplane from pitching down or up with changes in the center of gravity and center of lift, it needs to be able to trim. That's the role of the moving stabilizer. The stabilizer, that is, the entire horizontal tail, is hinged at the back, and the front end arcs up or down.

Here's how this is accomplished mechanically. Pushing the front end of the horizontal stabilizer up or down is done through a rotating jackscrew and a nut. The whole assembly works a bit like a carjack used to lift a vehicle, for example when changing a tire. You swivel, and the jackscrew rotates, pulling the so-called acme nuts inward and pushing the car up. In the MD-80 trim system, the front part of the horizontal stabilizer is connected to a nut that drives up and down a vertical jackscrew. An electrical trim motor rotates the jackscrew, which in turn drives the nut up or down. The nut then pushes the whole horizontal tail up or down.

On the 31st of January, 12 minutes after take-off, passing through 23,400 feet, the horizontal stabilizer had moved for the last time until the airplane's initial dive 2 hours and 20 minutes later.

Adequate lubrication is critical for the continued functioning of a jackscrew and nut assembly. Without enough grease, the constant grinding will wear out the thread on either the nut or the screw (in this case the screw is deliberately made of harder material, wearing the nut out first). The thread actually carries the entire load that is imposed on the vertical tail during flight. This is a load of around 5,000 pounds, similar to the weight of a whole family van hanging by the thread of a jackscrew and nut assembly. Were the thread to wear out on an MD-80, the nut would fail to catch the threads of the jackscrew. Aerodynamic forces then push the horizontal tailplane (and the nut) to its stop way out of the normal range, rendering the aircraft uncontrollable in the pitch axis, which is essentially what happened to Alaska 261. Even the stop failed because of the pressure. A so-called torque tube runs through the jackscrew in order to provide redundancy (instead of having two jackscrews, like in the preceding DC-8 model). But even the torque tube failed in Alaska 261.

On the surface, the accident seemed to fit a simple category: mechanical failure as a result of poor maintenance. A single component failed because people did not maintain it well. It had not been lubricated sufficiently. This led to the catastrophic failure of a single component. The break instantly rendered the aircraft uncontrollable and sent it plummeting into the Pacific.

But such accidents do not happen just because something suddenly breaks. There is supposed to be too much built-in protection against the effects of single failures. Other things have to fail too. More has to go wrong. And indeed, the investigation found more broken components in the system.

Closest to the flight crew, it found that there was no suggestion in any checklist that the flight crew should divert to the nearest possible airport when they got the first indications of horizontal stabilizer trouble. In fact, they found that the use of the autopilot with a jammed stabilizer was "inappropriate" and that a lack of guidance on how to fly an airplane with a jammed stabilizer could lead crews to experimenting and improvising, possibly making the situation worse.

As for the lubrication of the jackscrew, the investigation determined that the access panel in the tail plane of this aircraft type was really too small to adequately perform the lubrication task. Also, there had been widespread deficiencies in Alaska Airlines' maintenance program, leading, for example to a lack of adequate technical data to demonstrate that extensions of the lubrication interval would not present a hazard. There was also a lack of a task-by-task engineering analysis and justification in the process by which manufacturers revise recommended maintenance task intervals and by which airlines establish and revise these intervals.

Coupled to this, the investigation concluded that the end-play check interval (which measures how much play or slack there is in the screw/nut assembly) was inadequate. A restraining fixture for end-play checks was used even though it did not meet aircraft manufacturer specifications. In addition, the so-called on-wing end-play check procedure was never validated and was known to have low reliability. There was no requirement to record or inform customers that allowed overhauled jackscrew assemblies back onto airplanes that had higher end play than expected.

Finally, the investigation noted shortcomings in regulatory oversight by the Federal Aviation Administration, and an aircraft design that did not account for the loss of the acme nut threads as a catastrophic single-point failure mode.

This is the logic that has animated our understanding of failure for a long time now. Failure leads to failure. In order to explain a broken component (the jackscrew), we need to look for other broken components (the checklist, the access panel, the stress tests, the maintenance program, the company's oversight: all of them broken in their own way). In fact, any upbringing in Western science and engineering hardly allows people to think any other way. The first part of this book will explore the basis for this in Western thinking: the way in which our understanding of the world is rooted in the ideas of René Descartes and Isaac Newton, as well as many others who extended and popularized their ideas.

UNANSWERED QUESTIONS

But the idea that failure can be explained as looking for other failures leaves important questions unanswered. Questions, actually, that may stand between us and a firmer, further understanding of safety in complex systems. This is where the Newtonian–Cartesian basis of traditional human factors and systems safety is most limited, and limiting. Take, for example, the observation by the manufacturer that they knew of no reported cases of fatigue failures or fatigue damage on this torque tube design for their four basic models, the DC-9, MD-80, MD-90 and 717. No fatigue problems reported in over 95 million accumulated flight hours, and in over 2,300 airplanes delivered.[3] That would suggest that nothing was broken – neither in the parts, nor in the organizational processes surrounding their maintenance and approval.

A first question is quite simple. *Why*, in hindsight, do all these other parts (in the regulator, the manufacturer, the airline, the maintenance facility, the technician, the pilots) appear suddenly "broken" now? How is it that a maintenance program which, in concert with other programs like it never revealed any fatigue failures or fatigue damage after 95 million accumulated flight hours, suddenly became "deficient?" The broken parts are easy to discover once the rubble is strewn before our feet. But what exactly does that tell us about processes of erosion, of attrition of safety norms, of drift toward margins? What does the finding of broken parts tell us about what happened before all those broken parts were laid out before us? Our metaphors in safety, mostly driven by Newtonian–Cartesian thinking, can show us how things in an organization ended up. Not how they got there. The metaphors are mostly metaphors for resulting forms (for example, broken links between organizational compartments, layers of defense with holes), not models that capture processes of formation. There are few workable models that can even begin to capture the organic, jumbled, living interiors of the socio-technical organizations that do our risky business.

For more than 20 years now, safety researchers and practitioners have been talking about systems approaches, or systemic approaches. As a result, the discussions that ensued, the reports that got written, and models that were developed, have generally taken the context surrounding component failures more seriously. We now regularly look for sources of trouble in the organizational, administrative, and regulatory layers of the system, not just the operational or engineered sharp-end. The shift from sharp-end failures and operator error to the blunt end and organizational factors is certainly not complete, but at least legitimate today.

The Alaska investigation is testimony to this. It doesn't stop at the sharp end (either the cockpit or the maintenance shop). It goes all the way up through the blunt organizational end and beyond, to the regulatory authority. But it still returns with broken components. Which is the problem. That is not system thinking. System thinking is about relationships, not parts. System thinking is about the complexity of the whole, not the simplicity of carved-out bits. Systems thinking is about non-linearity and dynamics, not about linear cause-effect-cause sequences. Systems thinking is about accidents that are more than the sum of the broken parts. It is about understanding how accidents can happen when no parts are broken, or no parts are seen as broken.

Which produces a second question, perhaps even more fascinating. Why did none of these deficiencies strike anybody as deficiencies at the time? Or, if somebody did note them as deficiencies, then why was that voice apparently not sufficiently persuasive? If things really were as bad as we can make them look post-mortem, then why was everybody, including the regulator – tasked with public money to protect safety – happy with what was going on? Happy enough, in any case, to not intervene? You see, we create a huge problem for ourselves when we call these after-accident discoveries by all kinds of normative names, like deficiency, or inadequacy, or shortcoming. Inadequate or deficient or short relative to *what*? Clearly, people must have seen the norms that ruled their assessments and their decisions at the time as quite acceptable, otherwise pressure would have built for changing those norms. Which didn't happen.

These are questions at the heart of the future of safety. Behind the broken part onboard Alaska 261 lay a vast landscape of things that weren't all that broken, or not seen as broken at the time. There had been organizational trade-offs and decisions that all seemed quite normal, there was deregulation and increasing competitive pressure that was normal because it operated on every company. There were changes in regulations and oversight regimes. Which there always are. There were underspecified technical procedures and continuous quality developments in aircraft upkeep, which are normal because procedures are always underspecified and changes in how to conduct maintenance are always ongoing – including the routine extension of maintenance intervals. There had been collective international shifts in aircraft maintenance practices, following the kind of cross-national public-private initiatives that mark a global industry. And all this interacted with what one company, and its regulator, saw as sensible and safe.

One way to look at these influences, and the accident they eventually helped fashion, is as a story of drift: drift into failure. The story of drift is just that: a *story*, a way of telling things, of weaving them into a coherent narrative. The reason for telling it is not to claim a true account of what happened. The reason is to encourage a language, or a perspective on accidents, that is more open to the complexity, the dynamics, erosion and adaptation that marks socio-technical systems, than our thinking about accidents has been so far.

That is what this book invites us to do: to read a story of drift into the period before the accident or incident. To read a story of drift into what is going on inside an organization even without an incident or accident having taken place. Not because it is the correct angle to take – in fact, reading "drift" into what we see means making all kinds of analytical sacrifices. But we make analytical sacrifices no matter what perspective we take. No, the reason for looking for a story of drift is that it may help us stretch our imagination about safety and risk – beyond the components, beyond the now, beyond what's broken. Let's look at what such a story might tell us.

THE OUTLINES OF DRIFT

One of the greatest conundrums in safety work is to explain why a slide into disaster, so easy to see and depict in retrospect, was missed by those who inflicted it on themselves. Judging after the fact that people suffered a failure of foresight is easy. All we need to do is plot the numbers, trace out the trajectory, and spot the slide into disaster. But that is when we are standing amid the rubble and looking back. Then it is easy to marvel at how misguided or misinformed people must have been. But why was it that the conditions conducive to an accident were never acknowledged or acted on by those on the inside the system – those whose job it was to not have such accidents happen?

Foresight is not hindsight. There is a profound revision of insight that turns on the present. It converts a once vague, unlikely future into an immediate, certain past.[4] When we look back and we see drift into failure, it is, of course a byproduct of that hindsight. It seems as if the organization was inescapably heading toward the failure, and we almost automatically re-interpret the availability of signs pointing in that direction as really hard to miss. "People who know the outcome of a complex prior history of tangled, indeterminate events, remember that history as being much more determinant, leading 'inevitably' to the outcome they already knew"[5] But this certainty of trajectory and outcome is just ours. There was no trajectory and no outcome for the people involved at the time, or they would have done something about it. This is why I said that we can read a story of drift *into* the data, not out of the data. If we take a particular vantage point, that is. And standing in the aftermath of the failure definitely helps. The vantage point of the retrospective outsider is what helps us see that drift occurred.

But people's situations at the time would have looked very, very different. Sure, it may have offered multiple possible pathways. Some of these pathways or options might have been proven, some recognizable and familiar, and many of them plausible. But they would all have stretched out into the ever denser fog of futures not yet known. The decisions, trade-offs, preferences and priorities that people made, even if seemingly out of the ordinary and immoral after the failure, were once normal and common sense. Just as we like to believe that ours are now.

Drift, then, may be very difficult to recognize as drift *per se* without a bad outcome, without the benefit of hindsight. But there is probably another reason. It has to do with our thinking (and educating students) that complex systems are collections of components that all have different reliabilities. The "human component" is of course on the bottom reliability rung. It has to do with our belief that redundancy (through automation, procedures, back-up systems, monitoring) is the best way to protect against such unreliability.

Even before a bad outcome has occurred, indications of trouble are often chalked up to unreliable components that need revision, double-checking, more procedural padding, engineered redundancy. And this might all help. But it also blinds us to alternative readings. It deflects us from the longitudinal story behind those indications, it downplays our own role in rationalizing and normalizing signs of trouble. This focus on components gives us the soothing assurance that we can fix the broken parts, shore up the unreliable ones, or sit down together in a legitimate-sounding or expert-seeming committee that decides that the unreliability of one particular part is not all that critical because others are packed around to provide redundancy. And that all will be fine.

So not recognizing drift while it still happening, without an outcome to prove us right, could partly be a result of the language with which we have learned to look at our world. From whatever vantage point – retrospective or not – a different language will help us see different things. That is the point here. A lot has happened in various sciences since Descartes and Newton. There are interesting ideas and concepts that can offer us better insight into the dynamics, relationships, habituation, adaptive capacity, and complexity of the systems we want to understand and control. The previous chapter introduced five concepts that together may characterize drift:

- ○   Scarcity and competition
- ○   Decrementalism, or small steps
- ○   Sensitive dependence on initial conditions
- ○   Unruly technology
- ○   Contribution of the protective structure.

SCARCITY AND COMPETITION

The fuel, the energy, for adaptive organizational behavior (and, eventually, drift) is the simple fact that the organization is never alone. No organization operates in a vacuum. This is a fundamental feature of complexity and of complex systems: they are open to influences from the outside and are in constant transaction with what happens around them. Any organization operates and must try to survive in an environment that has real constraints, such as the amount of capital available, the number of customers reachable, and the qualifications of available employees. There are also hard constraints on how fast things can be built, developed, driven. And, most importantly, there are other organizations that try to do the same thing.

Many industries, aviation among them, are characterized by a dog-eats-dog business mentality in which any moment of weakness or inattention gets exploited immediately by a competitor. Deregulation set this in serious motion in the airline industry. Before the mid-1970s, the airline industry in most Western countries was essentially cartelized. Regulators were granted the authority by governments to assign certain routes to particular airlines, and to control the fares on these routes. There was a rationing of routes and players, resulting in strict limits on competition. New players had a very difficult time getting into the business, and rarely succeeded for long if they did. This arrangement kept airfares well above those that would be attained in a free market.

The 1978 Airline Deregulation Act in the U.S.A. changed all that, and was soon followed by similar initiatives in other Western countries. Market forces could now determine who could enter the industry, who could fly which routes and for what prices. The regulator at the time, the Civil Aeronautics Board (or CAB) was disbanded and replaced by the Federal Aviation Administration. Interestingly, deregulation became an asynchronous process in most countries: even though airlines were now allowed to compete under market forces, the infrastructure that they used remained in the hands of governments for a long time (airports, airways and the air traffic control system to run it), and even transnational arrangements often constrained heavily who could fly from which country to which. In the wake of deregulation, however, the airline industry expanded its employment by 32 percent, and passenger travel increased by 55 percent. The real cost of travel dropped by about 17 percent in the first decade after deregulation alone, and dropped even further in the ensuing decades.[6]

Deregulation cannot in itself be construed as a safety culprit, of course. Heavily regulated industries do not necessarily have a better safety record (if anything, it may encourage collusion between regulator and industry, particularly if part of the regulator's role is to encourage business development), and the period after deregulation actually saw a steady increase in airline safety.[7] But changing the rules of the game does change what goes on inside a complex system. And complexity theory can easily predict that changing the number of agents will change the dynamics of any complex system; it will affect the speed at which feedback about agents' actions travels and the patterns along which it reverberates. It might even change the way in which success is defined and assured, and also the way in which failure is bred and perhaps no longer recognized.

Jens Rasmussen suggested that work in complex systems is bounded by three types of constraints. There is an economic boundary, beyond which the system cannot sustain itself financially. Then there is a workload boundary, beyond which people or technologies can no longer perform the tasks they are supposed to. And there is a safety boundary, beyond which the system will functionally fail. For Rasmussen, these three constraints, or boundaries, surround the operation of any safety-critical system from three sides. There is no way out, there is only some room for maneuvering inside the space delineated by these three boundaries. Managerial and economic pressure for efficiency will push the system's operations closer to the workload and safety boundaries. Indeed, the likely result of increasing competitive pressure on a system, and of resource scarcity, will be a systematic migration toward workload and safety boundaries.[8]

Noncommercial enterprises experience resource scarcity and Rasmussen's three constraints too. With respect to the Alaska 261 accident, for example, a new regulatory inspection program, called the Air Transportation Oversight System (ATOS), was put into use in 1998 (two years prior to the accident). It drastically reduced the amount of time inspectors had for actual surveillance activities. A 1999 memo by a regulator field-office supervisor in Seattle suggested how this squeezed their oversight work into a corner where the safety and workload boundaries met:

> We are not able to properly meet the workload demands. Alaska Airlines has expressed continued concern over our inability to serve it in a timely manner. Some program approvals have been delayed or accomplished in a rushed manner at the "eleventh hour," and we anticipate this problem will intensify with time. Also, many enforcement investigations ... have been delayed as a result of resource shortages. [If the regulator] continues to operate with the existing limited number of airworthiness inspectors ... diminished surveillance is imminent and the risk of incidents or accidents at Alaska Airlines is heightened.[9]

Adapting to resource pressure, approvals were delayed or rushed, surveillance was reduced. Yet doing business under pressures of resource scarcity is normal: Scarcity and competition are part and parcel even of doing inspection work. Few regulators anywhere will ever claim that they have adequate time and personnel resources to carry out their mandates. Yet the fact that resource pressure is normal does not mean that it has no consequences. Of course the pressure finds ways out. Supervisors write memos, for example. Battles over resources are fought. Trade-offs are made. The pressure expresses itself in the common organizational, political wrangles over resources and primacy, in managerial preferences for certain activities and investments over others, and in almost all engineering and operational trade-offs between strength and cost, between efficiency and diligence. In fact, working successfully under pressures and resource constraints is a source of professional pride. Being able to create a program that putatively allows better inspections with fewer inspectors may win a civil servant compliments and chances at promotion, while the negative side effects of the program are felt primarily in some far-away field office.

Yet the major engine of drift hides somewhere in this conflict, in this tension between operating safely and operating at all, between building safely and building at all. This tension provides the energy behind the slow, steady disengagement of practice from earlier established norms or design constraints. This disengagement can eventually become drift into failure. As a system is taken into use, it learns, and as it learns, it adapts. A critical ingredient of this learning is the apparent insensitivity to mounting evidence that, from the position of retrospective outsider, could have shown how bad the judgments and decisions actually are. This is how it looks from the position of the retrospective outsider: the retrospective outsider sees a failure of foresight. From the inside, however, the abnormal is pretty normal, and making trade-offs in the direction of greater efficiency is nothing unusual. In making these trade-offs, however, there is a feedback imbalance. Information on whether a decision is cost-effective or efficient can be relatively easy to get. An early arrival time is measurable and has immediate, tangible benefits. How much is or was borrowed from safety in order to achieve that goal, however, is much more difficult to quantify and compare. If it was followed by a safe landing, apparently it must have been a safe decision. Extending a lubrication interval similarly saves immediately measurable time and money, while borrowing from the future of an apparently problem-free jackscrew assembly. Remember, the manufacturer knew of no reported fatigue problems or failures in 95 million flight hours accumulated by 2,300 airplanes.

Evidence from a feedback imbalance, then, suggests strongly that the system can operate equally safely, yet more efficiently. From the outside, such fine-tuning constitutes incremental experimentation in uncontrolled settings. On the inside, incremental nonconformity is an adaptive response to scarce resources and production goals. This means that departures from the norm become the norm. Seen from the inside people's own work, deviations become compliant behavior. They are compliant with the emerging, local ways to accommodate multiple goals important to the organization (maximizing capacity utilization but doing so safely; meeting technical or clinical requirements, but also deadlines).

As Weick pointed out, however, safety in those cases may not at all be the result of the decisions that were or were not made, but rather an underlying stochastic variation that hinges on a host of other factors, many not easily within the control of those who engage in the fine-tuning process.[10] Empirical success, in other words, is no proof of safety. Past success does not guarantee future safety. Borrowing more and more from safety may go well for a while, but we never know when we are going to hit. This moved Langewiesche to say that Murphy's law is wrong: everything that can go wrong usually goes right, and then we draw the wrong conclusion.[11]

DECREMENTALISM, OR SMALL STEPS

When it first launched the aircraft in the mid 1960s, Douglas recommended that operators lubricate the trim jackscrew assembly every 300 to 350 flight hours. For typical commercial usage, that could mean grounding the airplane for such maintenance every few weeks. Immediately, the sociotechnical, organizational systems surrounding the operation of the technology began to adapt, and set the system on its course to drift. Through a variety of changes and developments in maintenance guidance for the DC-9/MD-80 series aircraft, the lubrication interval was extended. These extensions were hardly the product of manufacturer recommendations alone, if at all. A much more complex and constantly evolving web of committees with representatives from regulators, manufacturers, subcontractors, and operators was at the heart of a fragmented, discontinuous development of maintenance standards, documents, and specifications. Rationality for maintenance-interval decisions was produced relatively locally, relying on incomplete, emerging information about what proved to be, for all its deceiving simplicity, unruly technology. Although each decision was locally rational, making sense for decision-makers in their time and place, the global picture became one of drift into failure.

Starting from a lubrication interval of 300 hours, the interval at the time of the Alaska 261 accident had moved up to 2,550 hours, almost an order of magnitude more. As is typical in the drift toward failure, this distance was not bridged in one leap. The slide was incremental: step by step, decision by decision that weaved through the aviation system from different angles and directions, alternating cross-industry initiatives with airline-driven ones, but all interacting to push the system into one direction: greater intervals.

In 1985, as an accompaniment of deregulation in the airline industry, jackscrew lubrication was to be accomplished every 700 hours, at every other so-called maintenance B check (which occurs every 350 flight hours).

In 1987, the B-check interval itself was increased to 500 flight hours for the entire industry, pushing lubrication intervals to 1,000 hours.

In 1988, B checks were eliminated altogether, and tasks to be accomplished were redistributed over A and C checks. The jackscrew assembly lubrication was to be done each eighth 125-hour A check: still every 1,000 flight hours.

But in 1991, A-check intervals were extended across the entire industry to 150 flight hours, leaving a lubrication every 1,200 hours. Three years later, the A-check interval was extended again, this time to 200 hours. Lubrication would now happen every 1,600 flight hours.

In 1996, Alaska airlines removed the jackscrew-assembly lubrication task from the A check and moved instead to a so-called task card that specified lubrication every 8 months. There was no longer an accompanying flight-hour limit. For Alaska Airlines, eight months translated to about 2,550 flight hours.

The jackscrew recovered from the ocean floor, however, revealed no evidence that there had been adequate lubrication at the previous interval at all. It might have been more than 5,000 hours since it had last received a coat of fresh grease. Through this drift, the effect of small things could suddenly explode into something huge. Miss one lubrication if you service the part every 350 hours, and you have no worries. Miss one when you lubricate every 2,550 hours, and you could get in deep trouble. This is where decrementalism, or small changes, can lead to big events. And where the way the system is organized (for example, around set lubrication intervals rather than continuous checks and lubrication-as-necessary) can make it highly sensitively dependent on such small changes.

SENSITIVE DEPENDENCY ON INITIAL CONDITIONS

This, in complexity and systems thinking, is called a sensitive dependency on initial conditions (or butterfly effect). Whether technology takes an unruly trajectory toward failure may depend on seemingly innocuous features or infinitesimally small differences in how it all got started.

The original DC-9 was certified in 1965 under Civil Aeronautics Regulations (CAR) 4b from three years prior (1962). The MD-80 was certified in 1980. More recent models of the DC-9, MD-80, MD-90, and Boeing 717 were certified under

14 CFR Part 25 and applicable amendments. However, systems that were similar to or that did not change significantly from the earlier DC-9 models, such as the longitudinal trim control system, were not required to be recertified. CAR 4b remained the certification basis for those parts of the MD-80, MD-90, and 717. This meant that even the most modern of the lineage, the Boeing 717, today has parts in it that are certified in 1965 under rules from 1962, and never needed to be re-certified. Just a time check: these rules were laid down a year before Kennedy was assassinated. When the first DC-9 was certified, we still had four years to go before landing on the moon. These were slide-rule times. No computer-aided design. The effects of, and sensitive dependency upon, such initial conditions are with us in lots of airplanes flying around today.

Understanding how a system may sensitively depend on initial conditions is something that certification processes are supposed to be able to do. But certification does not typically take lifetime wear of parts into account when judging (in this case) an aircraft airworthy, even if such wear will render an aircraft, like Alaska 261, quite unworthy of flying. Certification processes certainly do not know how to take socio-technical adaptation of new equipment, and the consequent potential for drift into failure, into account when looking at nascent technologies. Systemic adaptation or wear is not a criterion in certification decisions, nor is there a requirement to put in place an organization to prevent or cover for anticipated wear rates or pragmatic adaptation, or fine-tuning.

As a certification engineer from the regulator testified, "Wear is not considered as a mode of failure for either a system safety analysis or for structural considerations" (NTSB, 2002, p. 24). Because how do you take wear into account? How can you even predict with any accuracy how much wear will occur? McDonnell-Douglas surely had it wrong when it anticipated wear rates on the trim jackscrew assembly of its DC-9. Originally, the assembly was designed for a service life of 30,000 flight hours without any periodic inspections for wear. But within a year, excessive wear had been discovered nonetheless, prompting a reconsideration.

The problem of certifying a system as safe to use can become even more complicated if the system to be certified is sociotechnical and thereby even less calculable. What does wear mean when the system is sociotechnical rather than consisting of pieces of hardware? In both cases, safety certification should be a lifetime effort, not a still assessment of decomposed system status at the dawn of a nascent technology. Safety certification should be sensitive to the co-evolution of technology and its use, its adaptation. Using the growing knowledge base on technology and organizational failure, safety certification could aim for a better understanding of the ecology in which technology is released – the pressures, resource constraints, uncertainties, emerging uses, fine-tuning, and indeed lifetime wear.

Safety certification is not just about seeing whether components meet criteria, even if that is what it often practically boils down to. Safety certification is about anticipating the future. Safety certification is about bridging the gap between a piece of gleaming new technology in the hand now, and its adapted, coevolved, grimy, greased-down wear and use further down the line. But we are not very good at anticipating the future. Certification practices and techniques oriented toward assessing the standard of current components do not translate well into understanding total system behavior in the future. Making claims about the future, then, often hangs on things other than proving the worthiness of individual parts. Take the trim system of the DC-9 again.

The jackscrew in the trim assembly had been classified as a "structure" in the 1960s, leading to different certification requirements from when it would have been seen as a system. The same piece of hardware, in other words, could be looked at as two entirely different things: a system, or a structure. In being judged a structure, it did not have to undergo the required system safety analysis (which may, in the end, still not have picked up on the problem of wear and the risks it implied). The distinction, this partition of a single piece of hardware into different lexical labels, however, shows that airworthiness is not a rational product of engineering calculation. Certification can have much more to do with localized engineering judgments, with argument and persuasion, with discourse and renaming, with the translation of numbers into opinion, and opinion into numbers – all of it based on uncertain knowledge.

As a result, airworthiness is an artificially binary black-or-white verdict (a jet is either airworthy or it is not) that gets imposed on a very grey, vague, uncertain world – a world where the effects of releasing a new technology into actual operational life are surprisingly unpredictable and incalculable. Dichotomous, hard yes or no meets squishy reality and never quite gets a genuine grip. A jet that was judged airworthy, or certified as safe, may or may not be in actual fact. It may be a little bit unairworthy. Is it still airworthy with an end-play check of 0.0042 inches, the set limit? But "set" on the basis of what? Engineering judgment? Argument? Best guess? Calculations? What if a following end-play check is more favorable? The end-play check itself is not very reliable. The jet may be airworthy today, but no longer tomorrow (when the jackscrew snaps). But who would know? Nobody, really. Because the technology is unruly.

UNRULY TECHNOLOGY

Unruly means that something is disorderly and not amenable to discipline or control. It is a better word than uncertain. Very little to none of the technology we put into our complex systems is believed to be uncertain. Uncertain means unknown. And the whole point of engineering and validation and verification is to calculate our way out of such uncertainty, to not have such unknowns. We run the numbers, do the equations, build the computer simulations, so that we know under what load something will break, what the mean time between failures will be.

But despite all these efforts, there will always be unknowns, even if the technology is considered or judged or believed to be certain. The term "unruly technology" was introduced by Brian Wynne in 1988 to capture the gap between our image of tidiness and control over technology through design, certification, regulation, procedures, and maintenance on the one hand and the messy, not-so-governable interior of that technology as it behaves when released into a field of practice. Technology and safety is about how things behave in context, not on the drawing board. Universal proclamations or assurances about reliability figures are mute when ideas or designs are put into working solutions in this or that situation. A crucial skill involves finding a practical balance between universality of safety assumptions and their contextualization.[12]

This is a balance (and a possible gap) that not only operates between those on the outside (the public, or consumers of the technology) and insiders (engineers, managers, regulators), but applies even to insiders themselves: to practitioners very close to the technology. If the operational system is not itself following the rules by which it was predicted or supposed to operate, insiders can also reconcile such data with their beliefs – not by changing beliefs, but by looking differently at the data. This works particularly when (1) it is a relatively common problem, among a mass of other relatively common problems, (2) there are routine operational ways of compensating for it (providing more redundancy), and (3) alternative approaches would severely disrupt the economic or operational viability of the system.[13] The stretching of maintenance intervals, even against the background of a technology that behaves different in practice than what was predicted in the drawing, is a relatively common issue. Through these processes, insiders can keep their beliefs and they can retain their image of tidiness and controllability. But the technology may remain unruly, no matter what.

With as much lubrication of the jackscrew assembly as it originally recommended, Douglas thought it had no reason to worry about thread wear. So before 1967, the manufacturer provided or recommended no check of the wear of the jackscrew assembly. The trim system was supposed to accumulate 30,000 flight hours before it would need replacement. But operational experience revealed a different picture. After only a year of DC-9 flying, Douglas received reports of thread wear significantly in excess of what had been predicted. The technology, in other words, refused to play by the manufacturer's rules.

In response, the manufacturer recommended that operators perform a so-called end-play check on the jackscrew assembly at every maintenance C check, or every 3,600 flight hours. The end-play check uses a restraining fixture that puts pressure on the jackscrew assembly, simulating the aerodynamic load during normal flight. The amount of play between nut and screw, gauged in thousandths of an inch, can then be read off an instrument. The play is a direct measure of the amount of thread wear.

From 1985 onward, coinciding with the deregulation of the airline industry, end-play checks at Alaska Airlines became subject to the same kind of drift as the lubrication intervals. In 1985, end-play checks were scheduled every other C check, as the required C checks consistently came in around 2,500 hours, which was rather ahead of the recommended 3,600 flight hours. This would unnecessarily ground the aircraft for maintenance that was not officially due yet.

By scheduling an end-play test every other C check, though, the interval was extended to 5,000 hours. By 1988, C-check intervals themselves were extended to 13 months, with no accompanying flight-hour limit. End-play checks were now performed every 26 months, or about every 6,400 flight hours. In 1996, C-check intervals were extended once again, this time to 15 months. This stretched the flight hours between end-play tests to about 9,550.

The last end-play check of the accident airplane was conducted at the airline maintenance facility in Oakland, California in 1997. At that time, play between nut and screw was found to be exactly at the allowable limit of 0.040 inches. This introduced considerable uncertainty. With play at the allowable limit, what to do? Release the airplane and replace parts the next time, or replace the parts now? The rules were not clear. The so-called AOL 9–48A said that "jackscrew assemblies could remain in service as long as the end-play measurement remained within the tolerances (between 0.003 and 0.040 inch)."[14] It was still 0.040 inches, so the aircraft could technically remain in service. Or could it? How quickly would the thread wear from there on? Six days, several shift changes and another, more favorable end-play check later, the airplane was released. No parts were replaced: they were not even in stock in Oakland. The airplane "departed 0300 local time. So far so good," the graveyard shift turnover plan noted.[15]

Three years later, the trim system snapped and the aircraft disappeared into the ocean not far away. Between 2,500 and 9,550 hours there had been more drift toward failure. Again, each extension made local sense, and was only an increment away from the previously established norm. No rules were violated, no laws broken. Even the regulator concurred with the changes in end-play check intervals. These were normal people doing normal work around seemingly normal, simple, stable technology. Even the manufacturer had expressed no interest in seeing these numbers or the slow, steady degeneration they may have revealed. If there was drift, in other words, no institutional or organizational memory would know it. But isn't that exactly what we build protective structures for? Regulators, maintenance programs, accountable managers, nominated postholders?

CONTRIBUTION OF THE PROTECTIVE STRUCTURE

The operation of risky technology is surrounded by structures meant to keep it safe. The protection offered by these structures takes a lot of forms. In addition to the organization necessary to actually run the technology, there are regulatory arrangements, quality review boards, nominated post holders, safety departments, and a lot more. These deal with all kinds of information through meetings, committees, international or cross-industry collaborations, data monitoring technology, incident reporting systems, and a host of informal and formal contacts. All of which produces vast arrays of rules, routines, procedures, guidance material, prescriptions, expert opinions, engineering assessments and managerial decisions. And the people who populate this system change jobs and might even move from industry to regulator and back again. This is one of the features that makes complex systems complex: their boundaries are fuzzy. It is almost impossible to say where the system that makes the maintenance rules begins or ends.

There is, in other words, an astounding web of innumerable relationships in which the operation of risky technology is suspended – a web, moreover, that has no clear boundaries, no obvious end or beginning. This creates a wonderful paradox. The whole structure that is designed (and has evolved) to keep the technology safe, can make the functioning and malfunctioning of that technology more opaque. The meaning of signals about the technology (for example, that it is not behaving according to original manufacturer specifications, or that people are not abiding by the latest procedure) get constructed, negotiated, and transacted through the web of relationships that is strung throughout this structure. The weak signals that are left over trigger only weak organizational responses, if any at all.[16] What if the protective structure itself contributes to the construction and treatment of weak signals, and by extension to drift – in ways that are inadvertent, unforeseen, and hard to detect? The organized social complexity surrounding the technological operation, all the maintenance committees, working groups, regulatory interventions, approvals, and manufacturer inputs, that all intended to protect the system from breakdown, could actually help set its course to, and over, the edge of the envelope.

Take the lengthy, multiple processes by which maintenance guidance was produced for the DC-9 and later the MD-80 series aircraft. At first glance, the creation of maintenance guidance would seem a solved problem. You build a product, you get the regulator to certify it as safe to use, and then you tell the user how to maintain it in order to keep it safe. It was nowhere near a solved problem.

Alaska 261 illustrates the large gap between the production of a system and its operation. Inklings of that gap appeared in observations of jackscrew wear that was higher than what the manufacturer expected. Not long after the certification of the DC-9, people began work to try to bridge the gap. Assembling people from across the industry, a Maintenance Guidance Steering Group (MSG) was set up to develop guidance documentation for maintaining large transport aircraft, particularly the Boeing 747. Using this experience, another MSG developed a new guidance document in 1970, called MSG-2, which was intended to present a means for developing a maintenance program acceptable to the regulator, the operator, and the manufacturer.

The many discussions, negotiations, and inter-organizational collaborations underlying the development of an "acceptable maintenance program" showed that how to maintain a once-certified piece of complex technology was not at all a solved problem. In fact, it was very much an emerging understanding, open to constant revision and negotiation. Technology that appeared simple and certain on the drawing board, proved more unruly. It was not before it hit the field of practice that deficiencies became apparent, if one knew where to look.

In 1980, through combined efforts of the regulator, trade and industry groups and manufacturers of both aircraft and engines in the U.S.A. as well as Europe, a third guidance document was produced, called MSG-3. This document had to deconfound earlier confusions, for example, between "hard-time" maintenance, "on-condition" maintenance, "condition-monitoring" maintenance, and "overhaul" maintenance. Revisions to MSG-3 were issued in 1988 and 1993. The MSG guidance documents and their revisions were accepted by the regulators, and used by so-called Maintenance Review Boards (MRB) that convene to develop guidance for specific aircraft models.

A Maintenance Review Board, or MRB, does not write guidance itself, however; this is done by industry steering committees, often headed by a regulator. These committees in turn direct various working groups. Through all of this, so-called on-aircraft maintenance planning (OAMP) documents get produced, as well as generic task cards that outline specific maintenance jobs.

Both the lubrication interval and the end-play check for MD-80 trim jackscrews were the constantly changing products of these evolving webs of relationships between manufacturers, regulators, trade groups, and operators, who were operating off continuously renewed operational experience, and a perpetually incomplete knowledge base about the still-uncertain technology (remember, end-play test results were not recorded or tracked).

So what are the rules? What should the standards be? The introduction of a new piece of technology is followed by negotiation, by discovery, by the creation of new relationships and rationalities. "Technical systems turn into models for themselves," said Weingart, "the observation of their functioning, and especially their malfunctioning, on a real scale is required as a basis for further technical development."[17] Rules and standards do not exist as unequivocal, aboriginal markers against a tide of incoming operational data (and if they do, they are quickly proven useless or out of date). Rather, rules and standards are the constantly updated products of the processes of conciliation, of give and take, of the detection and rationalization of new data. As Brian Wynne said:

> Beneath a public image of rule-following behavior and the associated belief that accidents are due to deviation from those clear rules, experts are operating with far greater levels of ambiguity, needing to make expert judgments in less than clearly structured situations. The key point is that their judgments are not normally of a kind – how do we design, operate and maintain the system according to 'the' rules? Practices do not follow rules, rather, rules follow evolving practices.[18]

Nor is there a one-way and unproblematic relationship between the original rules or requirements and subsequent operational data. Even if the data, in one reading, may prove the original requirements or rules wrong, this doesn't mean that complex systems, under the various normal pressures of operating economically, reject the requirements and rules and come up with new ones. Instead, the meaning of the data can get renegotiated. People may want to wait for more data. The data can be denied. The data can be said to belong to a category that has nothing to do with safety but everything with normal operational variance (as happened in the *Columbia* Space Shuttle).

A STORY OF DRIFT

Setting up the various teams, working groups, and committees was a way of bridging the gap between building and maintaining a system, between producing it and operating it. Bridging the gap is about adaptation – adaptation to newly emerging data (for example, surprising wear rates) about unruly technology. But adaptation can mean drift. And drift can mean failure. Alaska 261 is about unruly technology, about gradual adaptations, about drift into failure. It is about the inseparable, mutual influences of mechanical and social worlds, and it highlights the inadequacy of our current models in human factors and system safety. Structuralist models are limited. Of course, we could claim that the lengthy lubrication interval and the unreliable end-play check were structural deficiencies. Were they holes in layers of defense? Sure, if we want to use that language. But such metaphors do not help us look for where the hole occurred, or why.

What kind of model could capture the gradual adaptation of a system like that surrounding jack screws, and predict its eventual collapse? There is something complexly organic about MSGs, something ecological, that is lost when we model them as a layer of defense with a hole in it; when we see them as a mere deficiency or a latent failure. When we see systems instead as internally plastic, as flexible, as organic, their functioning is controlled by dynamic relations and ecological adaptation, rather than by rigid mechanical structures. They also exhibit self-organization (from year to year, the makeup of MSGs was different) in response to environmental changes, and self-transcendence: the ability to reach out beyond currently known boundaries and learn, develop and perhaps improve.

Such an account needs to capture what happens within an organization, with the gathering of knowledge and creation of rationality within workgroups, once a technology gets fielded. A functional account could cover the organic organization of maintenance steering groups and committees, whose makeup, focus, problem definition, and understanding coevolved with emerging anomalies and growing knowledge about an uncertain technology. The journey that ended with the demise of Alaska 261 was marked by smalls steps, by a succession of decrements or downshifts in what people considered normal and acceptable, including those people (both internal and external to the company) tasked with safety oversight. The journey took place in a landscape of increasing competitive pressure and resource scarcity, constantly changing and evolving guidance on how to manage the unruly technology, and with normal people who came to work each day to do a normal job.

Can incident reporting not reveal a drift into failure? This would seem to be a natural role of incident reporting, but it is not so easy. The normalization that accompanies drift into failure (an end-play check every 9,550 hours is "normal," even approved by the regulator, no matter that the original interval was 2,500 hours) severely challenges the ability of insiders to define incidents. What is an incident? Before 1985, failing to perform an end-play check every 2,500 hours could be considered an incident, and given that the organization had a means for reporting it, it may even have been considered as such. But by 1996, the same deviance was normal, regulated even. By 1996, the same failure was no longer an incident.

And there was much more. Why report that lubricating the jackscrew assembly often has to be done at night, in the dark, outside the hanger, standing in the little basket of a lift truck at a soaring height above the ground, even in the rain? Why report that you, as a maintenance mechanic you have to fumble your way through two tiny access panels that hardly allow room for one human hand – let alone space for eyes to see what is going on inside and what needs to be lubricated – if that is what you have to do all the time? In maintenance, this is normal work, it is the type of activity required to get the job done. The mechanic responsible for the last lubrication of the accident airplane told investigators that he had taken to wearing a battery-operated head lamp during night lubrication tasks, so that he had his hands free and could see at least something. These things, such improvisations, are normal, they are not report-worthy. They do not qualify as incidents. Why report that the end-play checks are performed with one restraining fixture (the only one in the entire airline, fabricated in-house, nowhere near the manufacturer's specifications), if that is what you use every time you do an end-play check? Why report that end-play checks, either on the airplane or on the bench, generate widely varying measures, if that is what they do all the time, and if that is what maintenance work is often about? It is normal, it is not an incident.

Even if the airline had had a reporting culture, even if it had a learning culture, even if it had had a just culture so that people would feel secure in sending in their reports without fear of retribution, these would not be incidents that would turn up in the system. This is the banality of accidents thesis. These are not incidents. Incidents do not precede accidents. Normal work does. In these systems:

> accidents are different in nature from those occurring in safe systems: in this case accidents usually occur in the absence of any serious breakdown or even of any serious error. They result from a combination of factors, none of which can alone cause an accident, or even a serious incident; therefore these combinations remain difficult to detect and to recover using traditional safety analysis logic. For the same reason, reporting becomes less relevant in predicting major disasters.[19]

The failure to adequately see the part to be lubricated (that non-redundant, single-point, safety-critical part), the failure to adequately and reliably perform an end-play check – none of this appears in incident reports. But it is deemed "causal" or "contributory" in the accident report. The etiology of system accidents, then, may well be fundamentally different from that of incidents, hidden instead in the residual risks of doing normal business under normal pressures of scarcity and competition. This means that the so-called common-cause hypothesis (which holds that accidents and incidents have common causes and that incidents are qualitatively identical to accidents except for being just one step short of a true or complete failure) is probably wrong for complex systems.

The signals that are now seen as worthy or reporting, or the organizational decisions that are now seen as "bad decisions" (even though they seemed like perfectly good or reasonable proposals at the time) are seldom big, risky events or order-of-magnitude steps. Rather, there is a succession of weak signals and decisions, a long and steady progression of small, decremental steps of accepting them that unwittingly take an organization toward disaster. Each step away from the original norm that meets with empirical success (and with no obvious sacrifice of safety) is used as the next basis from which to depart just that little bit more again. It is this decrementalism that makes distinguishing the abnormal from the normal so difficult. Knowing what is worth reporting and what is not becomes lost in the fog of uncertainty and lack of differentiation. If the difference between what "should be done" (or what was done successfully yesterday) and what is done successfully today is minute, then this slight departure from an earlier established norm is not worth remarking or reporting on. Decrementalism is about continued normalization: It allows normalization and rationalizes it.

What we need, then, is not yet another structural account of the end-result of organizational deficiency. Not another story of broken components up and down the organizational ladder. What we need is a language that can help us get to a more functional account. A story of living processes, run by people with real jobs and real constraints but no ill intentions, whose constructions of meaning co-evolve relative to a set of environmental conditions, and who try to maintain a dynamic and reciprocal relation with their understanding of those conditions. This book intends to pry our logic away from hunting the broken components, and explore a language that might help us better understand the complex system behind the generation of success and failure.

REFERENCES

1   National Transportation Safety Board. (2000). *Factual Report: Aviation (DCA00MA023), Douglas MD-83, N963AS, Port Hueneme, CA, 31 January 2000.* Washington, DC: NTSB.

2   National Transportation Safety Board (2000). Ibid.

3   National Transportation Safety Board (2000). Ibid., p. 23.

4   Fischhoff, B. (1975). Hindsight is not foresight: The effect of outcome knowledge on judgment under uncertainty. *Journal of Experimental Psychology: Human Perception and Performance*, 1(3), 288–303.

5   Weick, K. (1995). *Sensemaking in organizations.* London: Sage, p. 28.

6   Poole, R.W. Jr., and Butler V. (1999). Airline deregulation: The unfinished revolution. *Regulation*, 22(1), 8.

7   Poole, R.W. Jr., and Butler V. (1999). Ibid.

8   Rasmussen, J. (1997). Risk management in a dynamic society: A modeling problem. *Safety Science*, 27(2/3), 183–213.

9   National Transportation Safety Board (2000). Ibid., p. 175.

10   Weick, K.E. (1993). The collapse of sensemaking in organizations: The Mann-gulch disaster. *Administrative Science Quarterly*, 38(4), 628–52.

11   Langewiesche, W. (1998). *Inside the sky: A meditation on flight.* New York, Pantheon Books.

12   Wynne, B. (1988). Unruly technology: Practical rules, impractical discourses and public understanding. *Social studies of science*, 18, 147–67.

13   Wynne, B. (1988). Ibid.

14   National Transportation Safety Board (2000). Ibid., p. 29

15   National Transportation Safety Board (2000). Ibid., p. 53.

16   Weick, K.E., and Sutcliffe, K.M. (2007). *Managing the unexpected: Resilient performance in an age of uncertainty*. San Francisco: Jossey-Bass.

17   Weingart, P. (1991). Large technical systems, real life experiments, and the legitimation trap of technology assessment: The contribution of science and technology to constituting risk perception. In T.R. LaPorte (ed.), *Social responses to large technial systems: Control or anticipation*, pp. 8–9. Amsterdam: Kluwer, p. 8

18   Wynne, B. (1988). Ibid., p. 153.

19   Amalberti, R. (2001). The paradoxes of almost totally safe transportation systems. *Safety Science*, 37, 109–26, p. 112.

3

THE LEGACY OF NEWTON
AND DESCARTES

Not long ago, a nurse from Wisconsin was charged with criminal "neglect of a patient causing great bodily harm" in the medication error-related death of a 16-year-old woman during labor. The nurse accidentally administered a bag of epidural analgesia instead of the intended penicillin by the intravenous route. The criminal complaint concluded that:[1]

○   The child's attending physician and the defendant's nurse supervisor reported that the nurse failed to obtain authorization to remove the lethal chemicals that caused the child's death from a locked storage system.

○   The nurse disregarded hospital protocol by failing to scan the bar code on the medication, a process of which the nurse had been fully trained and was cognizant. Had the lethal chemicals been scanned, medical professionals would have been forewarned of its lethality and the death would have been prevented.

○   The nurse disregarded a bright, clearly written warning on the bag containing the lethal chemicals prior to injecting them directly into the child's bloodstream.

○   The nurse injected the lethal chemicals into the bloodstream in a rapid fashion, failing to follow the approved rate for any medications that may have been prescribed for the child, in an apparent effort to save time. The rapid introduction of these chemicals dramatically hastened the death of the child, effectively thwarting any ability to save her life.

○   The nurse disregarded hospital protocol and failed to follow professional nursing procedures by not considering the five rights of patients prior to the administration of the lethal chemicals (right patient, right route, right dose, right time and right medication). The practice at her hospital requires the consideration of these five factors at least three times prior to the administration of any medication; the most important procedure established to prevent putting a patient's life in jeopardy through medication errors.

Interestingly, the charges were filed by an investigator for the Wisconsin Department of Justice, Medicaid Fraud Control Unit. The Fraud unit's job is to investigate and prosecute criminal offenses affecting the medical assistance program, which also includes anything that affects the health, safety and welfare of recipients of medical assistance. The teenage patient fell into that category, as, according to the charges, did the nurse's actions.

The investigator also reported going to the coroner's office and reviewing the epidural bag (that had contained the analgesic Bupivacaine) used in the incident. He found that the bag was labeled with an oversized, hot-pink label on the front side with bold black writing that states: CAUTION EPIDURAL. The bag had a second label on the backside that was hot-pink in color with bold black writing that stated FOR EPIDURAL ADMINISTRATION ONLY. The bag had a white label with a bar code for use with a computerized medication administration system. The penicillin bag, in contrast, did not have a pink cautionary label. It had two small orange labels, indicating the contents of the bag and the patient's name.[2]

The investigator's visit to the coroner's office, and looking at a silent evidence trail, was apparently not followed by a visit to the hospital to get any idea of the circumstances under which medication is administered. The nurse, for example, had volunteered to take an extra shift, because the hospital had difficulty providing nursing manpower, and she had had only a few hours of sleep during the previous 37 hours.

Now, as a result, the nurse, trying to work through the agony of having made a fatal error, faced potential action against her nursing license, and lost her job of 15 years. The criminal charges meant that she also faced the threat of 6 years in jail and a $25,000 fine. As the Institute for Safe Medication Practices wrote:

> It is important to keep in mind that there is usually much more to a medication error than what is presented in the media or a criminal complaint. For example, while the criminal complaint alleges that the nurse failed to follow the "five rights" and did not use an available bedside bar-coding system, some of the most safety-minded hospitals across the nation with bar-coding systems have yet to achieve a 100% scanning rate for patients and drug containers.

> This incident is similar to a 1998 case involving three nurses in Denver who were indicted for criminally negligent homicide and faced a possible 5-year jail term for their role in the death of a newborn who received IV penicillin G benzathine. At first glance, it appeared to many that disciplinary measures might be warranted in that case. But we found more than 50 deficiencies in the medication-use system that contributed to the error. Had even one of them been addressed before the incident, the error would not have happened or would not have reached the infant. Fortunately, in the Denver case, the nurse who stood trial was rightfully acquitted of the charges by a jury of laymen that deliberated for less than an hour.

> While there is considerable pressure from the public and the legal system to blame and punish individuals who make fatal errors, filing criminal charges against

a healthcare provider who is involved in a medication error is unquestionably egregious and may only serve to drive the reporting of errors underground. The belief that a medication error could lead to felony charges, steep fines, and a jail sentence can also have a chilling effect on the recruitment and retention of healthcare providers – particularly nurses, who are already in short supply.[3]

Newton's and Descartes' ideas have pretty much set the agenda for how we, in the West, think about science, about truth, about cause and effect. And how we think about accidents, about their causes, and what we should do to prevent them. Today these effects have become so ingrained, so subtle, so invisible, so transparent, so taken for granted, that we might not even be aware that much of the language we speak, and much of the thinking and work we do in safety and accident prevention, is modeled after their ideas. This is not only bad. Newton's and Descartes teachings have helped us shape and control our world in ways that would have been unfathomable in the time before they were around. Today we wouldn't be flying airplanes without Descartes. We wouldn't have put a space station in orbit without Newton. But when an airliner crashes on approach, or a Space Shuttle breaks and burns up on re-entry, or when the Piper Alpha oil platform is consumed by fire, or when a patient dies because an analgesic is accidentally pumped into her body via the intravenous line rather than the epidural one, we also call on Newton and Descartes to help us explain why things went wrong. And the consequences of doing that may not always be so helpful for making progress on safety.

The story that comes into view when Newton and Descartes help us explain a failure like this is pretty familiar. Consider the nurse's actions in the case above:

○   She should have known that such a dose of analgesic coursing into the body intravenously would pretty much lead to death. In fact, the investigator made her admit this on record. Newton could have told her so too. Newton loved the idea that the universe is predictable. As long as we know the starting conditions and the laws that govern it, which each responsible and competent nurse does for her or his little part of that universe, we can predict what the consequences will be.

○   The nurse may have lost situation awareness. That's Descartes. To Descartes, reality was a binary place. There's the world out there, and then there's the image of that world in our mind. But because we get tired or distracted or complacent (or, according to the complaint, negligent), we can "lose situation awareness" where the picture in our mind doesn't entirely line up anymore with reality, or isn't as complete. The "CAUTION EPIDURAL" label on the analgesic bag was very clearly available in the world, but somehow, because of criminal negligence, that caution failed to make it into the mind of the nurse.

○   The analgesic, when pumped in intravenously (and certainly when done as quickly as the criminal complaint asserts), represented an overdose. That's Newton. Newton was all about energy: the exchange of one form of energy for another (for example, chemical into physiological, as in this case. More chemical cause, more physiological effect. Like cardiac arrest, death). The Newtonian bastardization has become that if we want to contain danger, we need to contain energy. Or we have to carefully control its release (like we do in anything from jet engines to medication). The Swiss Cheese Model is about energy and containing it, too.

○   The cause for the patient's death was easy to find. The effect, after all, was the dead patient. And in Newton's world, causes happen before the effect, and ideally they happen pretty close to the effect. The nurse hooked up the wrong bag because of criminal negligence, and now the patient is dead. We found the cause.

○   Oh, and what about the nurse? She was fired and charged as a criminal. That's Newton, too. If there are really bad effects, there must have been really bad causes. A dead patient means a really bad nurse. Much worse than if the patient had survived. So much worse, she's got to be a criminal. Must be. We can't escape Newton even in our thinking about one of the most difficult areas of safety: accountability for the consequences of failure.[4]

So, if all this works so recognizably, then why should we bother to ask any critical questions of Newton and Descartes? Why doesn't their model of the world work just fine for our management of risk, for our safety improvements? Well, first, there's nothing inherently bad about having Newton and Descartes define how we should think. In fact, many of their ideas have really helped a lot in managing a complex, confusing world. But, like any model, theirs prevents us from seeing the world in other ways, from learning things about the world that we didn't even know how to ask.

WHY DID NEWTON AND DESCARTES HAVE SUCH AN IMPACT?

The worldview that lies at the basis of our thinking about systems, people, and safety, were formulated roughly between 1600 and 1750. This was roughly the period of what we now look back on as the Scientific Revolution. During that time, a wholesale shift occurred in how people related to the world around them, and to each other. A whole new way of thinking emerged, with features that we easily recognize because they are the basis for our own modern thought.

Society before 1600 was fundamentally different from what it would be a century later, and of course from what it is today. In medieval times, people lived in small, cohesive communities, characterized by interrelationships between people and nature.[5] During the Scientific Revolution, and its concomitant growth of trade, industry, knowledge exchange, travel and urbanization, this was replaced by a so-called mechanistic worldview. "Mechanistic" is a keyword. It would become the way of looking at the world, at relationships, at systems. The achievements of Copernicus, Galileo, Kepler and later indeed Newton during the Scientific Revolution meant that people came to regard the world as a machine, as something consisting of parts and interactions between those parts, governed by laws that could be captured in mathematical formulas.

Galileo Galilei, Italian physicist, mathematician and astronomer (1564–1642), promised people that they can actually understand the world, that they can discover regularities, experiment with them, and describe the results using a mathematical language. Knowledge of the world became a matter of putting observations in geometrical figures and equations. Experimentation and mathematical description of nature, Galileo's contribution, were to become the two dominant features of the scientific enterprise from the seventeenth century onward. Galileo, who died a year before Newton was born, insisted that we should not worry too much about the things we can't quantify (like how something tastes or smells) because that's a subjective mental projection and not a material property of the world that can be measured. And since you can't measure it, it's not very scientific. That legacy lives with us still today: an obsession with measurement and quantification and a downplaying of the things that not only help determine the outcome of many phenomena (like our values, feelings, motives, intentions) but that also make us human.

At the time, the Galilean project was full of promise, as the world had not been an easy or very manageable place before. Medieval people were at the mercy of whatever nature threw at them. The Black Death (or Great Plague) peaked around 1350, reducing Europe's population by between 30 percent and 60 percent by 1400. It would take until the mid-1500s to recover to a pre-pandemic level, leaving in its wake religious, social and economic upheavals and the stinging realization that any prediction of, or control over, what could happen to people was very difficult, if not impossible. The goal of science, of assembling knowledge, English statesman and philosopher Francis Bacon (1561–1626) decided, was to dominate and control nature. Science should "torture nature's secrets from her," he wrote. Nature had to be "hounded in her wanderings," "bound into service," "put into constraint" and made a "slave." With nature having been the unpredictable force in people's lives for millennia, and people simply at its uncontrollable, receiving end, such talk must have been an elixir, a welcome relief. Seeing nature as a machine was liberating.

Consider for a moment the depth of that relief. A French peasant, born in the eighteenth century, had a rough chance of one in four of surviving infancy, and then a life expectancy of about 22 years. That is about the same as for a comparable Roman citizen almost two millennia before. Women in Europe lived even shorter lives than men; married life was short, and many men remarried as widowers. During Bacon's and Newton's lifetimes, Italy, Spain, Germany, Austria, France, Russia and England all had severe outbreaks of epidemic diseases, with an estimated 100,000 people killed in London alone in 1665–1666, paring the city's population by 20 percent. Famine remained a real threat, too. There was a chronic fear that soon there would not be enough to eat, and people generally had no options other than pursuing the way of life as it had been handed down to them.

What we now call the scientific revolution helped change this outlook. Francis Bacon pursued the experimental scientific method with zest in England, while Galileo was experimenting and deflecting the Church in Italy. Bacon was the first to articulate induction: the ability to draw generalizable conclusions from specific, limited experimental encounters with the world. Agriculture offered the first living laboratory of what could be done by rudimentary science – systematic experimenting, observing, recording – to improve people's control over their environment in ways that certainly beat sticking to custom or lore. Technological innovations came next. Husbandry improved, as did crop yields, necessary to sustain the continuing rise of a European population. Innovations in commerce and agriculture depended on a steady stream of specialists, and many universities were established that educated whole classes of bureaucrats and engineers in forestry, mining, veterinary medicine and agriculture. Their output appealed to governments attempting to exert more control than ever over their natural environments and reap the resources on which they depended.[6]

Newton (1643–1727) was the one who would supply the laws that nature supposedly adhered to. Three really important ones, in fact. If you know the laws (and the starting conditions to plug into them), Newton said that you can foresee or predict how nature is going to behave. And then you can perhaps do something about it. Again, the prospect of control over nature made great sense against the background of having struggled through a couple of millennia without much of that control. Newton was religious and asserted that it was God who had created nature. This not only delighted the Church, of course, for it kept science and religion in close harmony and put the Church front-and-center, it was also consistent with how people had viewed the world up to that point: as God's creation. But what Newton now promised was that it was possible for humans to uncover its hidden laws. Perhaps that was even humanity's sacred duty, to figure out, bit by bit, how God had designed nature, how God had put it all together. Since we couldn't really dial up and ask how He'd done it, people had to discover the laws experimentally, using the scientific method, for which Newton himself had written the recipe in his most famous 1687 book, commonly known as the *Principia*.[7]

When Newton was admitted to Cambridge in 1661, he became familiar with the writings of René Descartes (1596–1650), French mathematician and philosopher. The intellectual encounter became deeply influential, so much so that today we have no trouble lobbing the two names together in a label like the "Newtonian–Cartesian worldview."[8] Descartes is considered the first (and a formidable) modern philosopher. Like Galileo, Descartes believed that the riddle of nature was written in mathematical laws: equations and geometries could be deduced, with which the workings of the world could then be captured, or described. Descartes had concluded that the absolute truth about the world could thus be discovered. Scientific knowledge, if arrived at by the right method, leads to absolute certainty. His introduction to science was published in his *Discourse on Method* in 1637. Its full title held all the promise of Cartesian certainty: it would take the reader through *the Method of Rightly Conducting One's Reason and Searching the Truth in the Sciences.* The truth about the world was available to those following the method rightly.

Both Newton and Descartes dealt with complex phenomena through a technique called reductionism. In order to understand a complex phenomenon, you have to take it apart so as to discover the components that make it up. If those components are still too complex to understand, you take them apart too. And you go on, until you reach the simplest level. That's how you make complexity understandable. And that's how you can assert control over complex phenomena. This itself was an active departure from classical Greek philosophy, such as Aristotle's theories on matter (made of four elements: fire, earth, air and water). Irish physicist and chemist Robert Boyle (1627–1691) proposed, not unlike Newton, that the universe could be explained from three original principles: matter, motion and rest. Matter itself was capable of reduction to minute particles (*minima*, or *prima naturalia*, which Boyle called "single" and "sensible") which, in coalition or cluster, made up the physical and chemical world. Reductionism has been the mainstay of the scientific method for hundreds of years, and, from its origins in physics and chemistry, has been the analytic driving force behind everything from engineering to biology to medicine and psychology. In order to understand and influence phenomena, you have to go deeper and deeper, understanding what the constituent components are and how they interact. Reductionism has had an ever-lasting impact on our thinking about systems and safety. Its impact is understandable for the promise it holds. The secrets of complexity can be "tortured out of" systems by breaking them up into smaller, more manageable parts.

Another central idea, proposed by Descartes, was that of dualism. In trying to explain how we can be aware of the world, Descartes proposed that we have a mind that is separate from the material world, and that these two areas shared no common features, and could not be explained in the same way. Even though both were the creation of God, the world *had* to be explained, whereas the mind *did* the explaining. The material universe, to Descartes, was a machine. It was a clever machine, for sure, that much God could be counted on. But still a machine that behaved according to the laws and rules that govern machines. If you could explain the arrangement and movement of parts, you could understand it. This went for biological systems too, Descartes argued. Animals were also machines, and could be modeled like clockwork, which had attained a high degree of perfection in Descartes' time. All natural phenomena could be accounted for in one single system of mechanical principles. Newton would later supply the mathematical laws that did exactly that. This mechanical picture of nature would become dominant in scientific and cultural thinking for centuries to come, as would the idea of a separate mind whose job it was to understand that world. While seeing nature as a machine was once liberating, it may now have become imprisoning.

In the following chapters, I will flesh out better what consequences these ideas have had for our thinking about systems and safety, in particular how we think about the role of humans and human work in safety-critical systems, how we conceptualize risk and how to contain it, and how we think about accountability.

But first, a brief fast-forward of some problems of sticking with a Newtonian–Cartesian worldview when thinking about safety.

SO WHY SHOULD WE CARE?

At the very least, the risk of the Newtonian–Cartesian worldview lies in us not realizing that theirs is just one way of looking at the world, one way of putting things. One language. One set of concepts and assumptions. The risk lies in our not realizing that these assumptions have become transformed into realities that we take for granted in our thinking about the world, and in how we believe we should manage it. This risk means that we are not very reflective, not very critical, about our own assumptions, about our own ideas. It means that we are excluding other perspectives – not because we have evaluated them and decided they are useless, but because we don't even recognize that they exist. For all the wisdom given to us by Descartes and Newton, it makes us really dumb if we assume that theirs is the only way to think about the world.

Go back to the nurse. Of course we can use Newton to say that there is a cause (the bag was hooked up wrongly) and an effect (a dead patient). But is that really the whole story? Or the only story? Developments in safety thinking have pushed us, over the past 20 years, to look beyond that so-called "proximal" cause. It has told us that we need to look for the organizational "causes" of the failure too. These have, in the Swiss Cheese Model, been called the "latent" causes, or latent factors. It has been popular to believe that this development has brought us systemic thinking about safety, about accidents. But it hasn't. Not really. It's still very Newtonian. The models that have popularized this notion suggest that the organizational causes of accidents include inadequate supervision, deficient line management, imperfect defenses, unsafe acts. These are among the components that make up the success or failure of a complex system – all lined up in a causal sequence. That is Newtonian.

We are still looking for causes, and for the effects of those causes. We are still looking for broken components, even if we search higher up the organizational ladder. We are still assuming that people inside an organization have the ability to form an adequate picture of the world around them but that they fail to do so because of communication difficulties, hubris, complacency, or other cognitive limits. That's all Descartes and Newton. Nothing fundamental has changed in how we view the world. Models of organizational safety over the past 20 years have only asked us to stretch up our imagination a bit, to enlarge that world by a few levels, that's all. Now, to be fair, that in itself is a huge accomplishment. In societies that prize easy answers to complex accidents, that love to have experts find "the cause" of the failure, it has been an important addition to the vocabulary. But it has changed nothing about our basic assumptions of safety and risk.

WHAT NEWTON CAN AND CANNOT DO

Newton was of great help in understanding how the Space Shuttle *Columbia* came apart during its re-entry into the earth's atmosphere on 1 February, 2003.

Mechanistic thinking about causes and effects located the triggering event at 81.7 seconds into the launch 16 days earlier. A large piece of hand-crafted insulating foam came off an area where the Shuttle attaches to the external tank, and struck the leading edge of *Columbia*'s left wing at 81.9 seconds when the Shuttle was at 65,600 feet altitude and climbing at two and a half times the speed of sound. *Columbia* re-entered the earth's atmosphere 16 days later with a pre-existing breach in the leading edge of its left wing in the vicinity of reinforced carbon-carbon (RCC) panel 8.

This breach, caused by the foam strike on ascent, was of sufficient size to allow superheated air (probably exceeding 5,000 degrees Fahrenheit) to penetrate the cavity behind the RCC panel. The breach widened, destroying the insulation protecting the wing's leading edge support structure, and the superheated air eventually melted the thin aluminum wing spar. Once in the interior, the superheated air began to destroy the left wing. The Shuttle continued to fly its pre-planned flight profile, although, still unknown to anyone on the ground or aboard *Columbia*, her control systems were working furiously to maintain that flight profile. Finally, over Texas, increasing aerodynamic forces in the denser levels of the atmosphere overcame the catastrophically damaged left wing, causing the *Orbiter* to fall out of control at speeds in excess of 10,000 mph.

Interestingly, there was plenty of data to carefully reconstruct the Newtonian burning-and-breaking sequence upon re-entry. Recordings from hundreds of sensors inside the wing, and analyses of the reactions of the flight control systems to the changes in aerodynamic forces, allowed NASA to not only follow (though quite disbelievingly) the breakup as it happened, but also to accurately recreate the mechanistic order in which things interacted and went terribly wrong. Going down and in, relying on the narrow, precise perspectives from the multitude of sensors that offered a view from inside the wing, deep inside the system, it was possible to lay bare the Newtonian exchanges of energy, the rearrangements of particles within Euclidean space.

COLUMBIA: PARTS THAT BROKE IN SEQUENCE

The *Columbia* accident investigation details the sequence of discoveries about the parts that failed. While losing communication with the Space Shuttle, it was thought that the failures were due to instrumentation problems, not actual part failures. At 8:54:24 a.m., the Maintenance, Mechanical, and Crew Systems (MMACS) officer and the Flight Director (Flight) started the following exchange:

MMACS: "Flight – MMACS."
Flight: "Go ahead, MMACS."
MMACS: "FYI, I've just lost four separate temperature transducers on the left side of the vehicle, hydraulic return temperatures. Two of them on system one and one in each of systems two and three."
Flight: "Four hyd [hydraulic] return temps?"
MMACS: "To the left outboard and left inboard elevon."

Flight: "Okay, is there anything common to them? DSC [discrete signal conditioner] or MDM [multiplexer-demultiplexer] or anything? I mean, you're telling me you lost them all at exactly the same time?"
MMACS: "No, not exactly. They were within probably four or five seconds of each other."
Flight: "Okay, where are those, where is that instrumentation located?"
MMACS: "All four of them are located in the aft part of the left wing, right in front of the elevons, elevon actuators. And there is no commonality."
Flight: "No commonality."

At 8:56:02 a.m., the conversation between the Flight Director and the MMACS officer continues:

Flight: "MMACS, tell me again which systems they're for."
MMACS: "That's all three hydraulic systems. It's two of them are to the left outboard elevon and two of them to the left inboard."
Flight: "Okay, I got you."

At 8:58:00 a.m., *Columbia* crossed the New Mexico–Texas border:

MMACS: "Flight – MMACS."
Flight: "Go."
MMACS: "We just lost tire pressure on the left outboard and left inboard, both tires."

At 9:00:18 a.m., the postflight video and imagery analyses indicate that a catastrophic event occurred. Bright flashes suddenly enveloped the Orbiter, followed by a dramatic change in the trail of superheated air. This is considered the most likely time of the main breakup of *Columbia*:

MMACS: "Flight – MMACS."
Flight: "MMACS?"
MMACS: "On the tire pressures, we did see them go erratic for a little bit before they went away, so I do believe it's instrumentation."
Flight: "Okay."

In Mission Control, there was no way to know the exact cause of the failed sensor measurements, and while there was concern for the extended loss of signal, the recourse was to continue to try to regain communications and in the meantime determine if the other systems, based on the last valid data, continued to appear as expected. At 9:12:39 a.m., however, *Columbia* should have been banking on the heading alignment cone to line up on Runway 33. At about this time, a member of the Mission Control team received a call on his cell phone from someone who had just seen live television coverage of *Columbia* breaking up during re-entry. The Mission Control team member walked to the

Flight Director's console and told him the Orbiter had disintegrated. Flight then calls the Ground Control Officer (GC).

Flight: "GC, – Flight. GC – Flight?"
GC: "Flight – GC."
Flight: "Lock the doors"

Having confirmed the loss of *Columbia*, the Entry Flight Director directed the Flight Control Team to begin contingency procedures. The exchanges between the various people involved in the unfolding drama impart very important Newtonian–Cartesian features of how we understand accidents. A first feature is that of a sequence. One thing happens after another. One thing causes another, and leads to another. This spreads out over time; time is the organizing principle. A second is that of data and measurements and instruments. An accurate understanding requires accurate data, which requires accurate instrumentation. The role of science is in part to develop better instruments, so as to get better data, so as to get a better understanding.

When MMACS told Flight, "I do believe it's instrumentation," the message was that the data was so unbelievable that it had to be an instrumentation error: the data reported by the instruments could not be from some real world out there. That the Space Shuttle would be breaking up was not conceivable. So a request was made for more data. "Any other trackers we can go to?" It was only when the team got a cell phone call, reporting live footage of the *Columbia* breaking up, that their notion of an instrumentation problem collapses. Better data, more "truthful" data, through live television footage, trumped the hundreds of data sources NASA itself had fixed into the Shuttle. When their signals were lost, NASA's knowledge of what was going on was lost.

The Newtonian idea is that if you can't find out what's going on, you have to get a better instrument. If it's too far, you get a telescope. If it's too small, you get a microscope. If it's too remote, you get a boroscope or an endoscope. Then you get better data, and then you get a more accurate picture. In this case, that better picture was brought by a television camera. This notion, that data gathered about a break-up represents something very "real" (and that some data is more "real" than other data) about the accident, about how it happened, is firmly rooted in the Newtonian–Cartesian vision of what knowledge is. Descartes and Newton had pretty strong ideas on what such knowledge represents, and what role that knowledge plays in our understanding of the world and the accidents that happen in it. More on this in later chapters.

BROKEN PART, BROKEN SYSTEM?

What such a sequence and the following analysis could not show, however, was how cultural traits and organizational practices related to safety had developed or stayed in place, even as the 1986 Space Shuttle *Challenger* accident echoed eerily through the social, political and organizational stories behind the *Columbia* breakup. This included a seeming reliance on past success as a substitute for what could be considered to be sound engineering practices (like testing to understand why systems were not performing in accordance with requirements). Also, features of the bureaucracy's and contractors' organizational structures deflected and parried and narrowed what turned out to be critical safety information and stifled professional differences of opinion. Other structural features that interacted with organizational processes were a fragmentation of management across program elements, and the evolution of informal relationships and decision-making processes that operated outside the formal organization.

These processes jointly helped the organization meander and drift slowly from one accident to another inside two decades. And, in contrast to the hundreds of sensors in the Space Shuttle's wing alone, there were no hundreds of sensors across NASA and its contractors that can send "accurate" data on the influence of politics, budgets or engineering decisions on the safety of the next launch or re-entry. To explain the structure and processes inside the NASA organization, the *Columbia* accident board did not go down and in – it went up and out.

Its predecessor, the Presidential Commission set up to investigate the *Challenger* accident, did not really go up and out. In important parts of its report, it went down and in, in faithful Newtonian fashion. A Space Shuttle broke up during the launch. What were the broken parts inside the system that could explain that break-up at system level? Recall from the beginning of the book how the Presidential Commission told NASA to correct its flawed decision-making, calling for adaptations of individual behavior and procedures. The tendency of engineers and managers not to report upward needed to be corrected through changes in personnel, organization, or indoctrination. The report placed responsibility for "communication failures" not with the organizational structure, but with individual middle managers, who had overseen key decisions and inadequate rules and procedures.[9]

The step to saying that failed parts had been found, in good Newtonian fashion, was not very large. And of course, one is forced to wonder (though we will probably never know) what role this identification of a few eureka parts (and particularly how it may have excluded other stories, other voices), may have contributed to an accident with a similar organizational silhouette 17 years later. Richard Feynman, physicist and maverick member of the Presidential Commission that investigated the 1986 *Challenger* accident, wrote a separate, dissenting appendix to the report on his own discoveries. One that disturbed him particularly was the extent of structural disconnects across NASA and contractor organizations, which uncoupled engineers from operational decision-making and led to an inability to convince managers of (even the utility of) an understanding of fundamental physical and safety concepts.[10] But it was the deviating voice, the dissenting one, not the mainstream legitimated one.

With the remnants of two broken Space Shuttles to look at, the *Columbia* accident investigation tried to understand how structural properties of NASA and its contractors influenced what was, or could be, seen as rational and safe by those responsible. The investigation attempted to locate these properties in the much larger configuration of societal, historical, political forces that all influenced and shaped them. It did not try to locate them in the traits and mindsets of individual managers or engineers. How did history and culture contribute? What were the influences of fluctuating competitive pressures in the environment? The *Columbia* accident investigation tried to trace how the original compromises that were required to gain approval for the Shuttle interacted with subsequent decades of resource constraints, fluctuating priorities of successive federal administrations, schedule pressures, the characterization of the Shuttle as operational rather than developmental, and the lack of an agreed national vision for human space flight. Those are hardly the Newtonian parts that can be pinpointed as having failed.

Newtonian logic, language and laws can nicely capture the physical causes of the *Columbia* accident. Newtonian assumptions about the symmetry between cause and consequence, as well as the preservation of energy through successive conversions – gravity, speed, friction, heat, breakup – can meaningfully be used for modeling important aspects of the last few minutes of the Shuttle's re-entry. It even offers us the sort of linear depiction of events that could help with the identification of barriers that help in future prevention. For example, things NASA or its subcontractors could do would be better pre-entry inspection of damage to hull or wings that could promote unwanted energy transfer or release, to separating people from the energy being released by creating a crew escape module.

ACCIDENTS COME FROM RELATIONSHIPS, NOT PARTS

In its success at explaining the physical cause of the *Columbia* accident also lies the risk of a retaining Newtonian–Cartesian worldview. This worldview could actually be actively hostile to our abilities to make further progress on safety. The focus on sequences and parts has probably contributed to the idea that the best way to battle risk in such systems is to put in more barriers, to create greater redundancy. Barriers can stop progressions to failure, redundancy can make a whole part substitute for a broken one, or take a system off a pathway to failure and direct its progression towards a more successful outcome. That's the idea. As explained in the second part of the book, however, this idea comes at a huge cost. More redundancy, which itself comes from this fixation with parts, creates greater opacity, greater complexity. The system can become more difficult to manage, impossible to understand. With the introduction of each new part or layer, there is an explosion of new relationships (*between* parts and layers and components) that spreads out through the system. The insight of the late twentieth century is that accidents emerge from these relationships, not from some broken parts that lie in between.

Not only does a Newtonian approach itself contribute to greater system complexity, its ability to explain such complexity is very limited. Understanding and dealing with unwanted Newtonian exchanges and releases of energy in a linear, closed system is one thing. And, if you believe the *Columbia* Accident Investigation, relatively simple. Doing the same thing for a social-technical organization that was constructed to build and operate the technology safely is something entirely different. Newtonian logic is not helpful for meaningfully modeling the messy, constantly changing, kaleidoscopic interiors of organizational life where failures and successes are spawned. Linear sequences of causes and effects, where big causes account for big effects, can never fairly capture the workings of a NASA and its subcontractors where lots of little things eventually set something large in motion, where there were thousands of little assessments and daily actions from a huge, distributed workforce, in which everybody came to work to try to do a good job, where everybody was dealing with their own daily innumerable kinds of production pressures, their PowerPoint presentations and seemingly innocuous decisions about maintenance and flight safety.

From Newton and Descartes, we get no help at all for seeing, let alone understanding or modeling, the very subtle processes of normalization, of an incremental acceptance of risk, of gradual adaptation that are endemic to living organisms, from people to huge organizations. These kinds of processes remain opaque to the Newtonian–Cartesian perspective. Or, to put it another way, the Newtonian–Cartesian perspective finds answers to questions that are not so interesting, like being able to show post-mortem where the holes in the organization's layers of defense were. This is not just an academic concern. If Newton fails to deliver us a cause (and if we have no other approach to understanding how to intervene), we have nothing to fix. And if we have nothing to fix, the failure can happen again. As indeed it might have, as far as Space Shuttles were concerned.

Of course, we can try to follow the Newtonian cause-effect trail back from the break-up and burn-up of the Space Shuttle *Columbia*. But the further away we move in time and space from its debris trail over the Southern U.S.A., the more our attempts to map causal pathways get stranded. The further we move away, the less things look broken. The more "normal" everything looks. So where did the eventual failure originate, really? We end up in the squishy, intractable morass of normal people working unremarkably in a U.S. government bureaucracy, subject to normal tug-of-war budget problems, career moves, congressional politics, societal expectations and the ghost of post-Sputnik history. All quite normal, really. In that morass, the Newtonian trail goes cold.

Not being able to come up with a meaningful account of what happens inside the dynamic political, bureaucratic and other social processes that govern a complex organization like NASA, and how it interfaces with engineering logic and knowledge, represents a serious shortcoming in safety improvement work. As long as such work, deep down, remains wedded to Newtonian–Cartesian assumptions about how the world works, we will lose a lot, miss a lot, and repeat the logic that may have got the system in trouble before. As an example, the recommendations from the Presidential Commission tasked with investigating the 1986 *Challenger* accident were relatively componential. NASA had to correct its flawed decision-making, and the report called for changes in individual behavior and procedures. The tendency of managers not to report upward needed to be corrected through changes in personnel, organization, or indoctrination (indeed, through changes in three kinds of components). Ultimately, the report placed responsibility for "communication failures" not with the organizational structure, but with individual middle managers, who had overseen key decisions and inadequate rules and procedures.[11]

The problem of creating interventions into cultural, political, organizational (or any other "social") factors with a sustained Newtonian–Cartesian worldview was expressed at a symposium at the Massachusetts Institute of Technology: "Public reports following both shuttle disasters pointed to what were termed organizational and safety culture issues, but more work is needed if leaders at NASA or other organizations are to be able to effectively address these issues ... systematically taking into account social systems in the context of complex, engineered technical systems. Without a nuanced appreciation of what engineers know and how they know it, paired with a comprehensive and nuanced treatment of social systems, it is impossible to expect that they will incorporate a systems perspective in their work."[12] The point of the second part of this book is to explore the landscape of ideas and concepts on offer that could help us develop such a more "nuanced treatment of social systems" in our safety improvement efforts.

Newton and Descartes both died a long time ago. But their legacy is quite alive. The issue is not that the Newton's and Descartes' way of understanding the world is necessarily bad. Rather, the issue is that each position, each perspective, makes analytical sacrifices. The sacrifices that we make when seeing the world through a Newtonian–Cartesian lens could lead to us shortchanging ourselves. They could retard the development of a deeper understanding of why things might go right and wrong in complex systems.

WE HAVE NEWTON ON A RETAINER

Let's go back to the case from the beginning of this chapter, where a 16-year-old patient died after a nurse accidentally administered a bag of epidural analgesia by the intravenous route instead of the intended penicillin. The dominant logic of Newton and Descartes helped turn the nurse into the central culprit, leading to her criminal conviction. But the influence of Newton and Descartes did not stop there. A root cause analysis was done (more typically known as an RCA in the healthcare world). RCA's are designed to track from the proximal events to the distal causes, or indeed the root causes. The idea of a root cause, of course, is very Newtonian. Effects cannot occur without a cause that we can trace back and nail down somewhere. And that track has a definitive, determinate end: the root cause or causes that triggered all subsequent events.

The RCA identified four proximate causes of the nurse's error. These were the availability of an epidural medication in the patient's room before it was prescribed or needed, the selection of the wrong medication from a table, the failure to place an identification band on the patient, which was required to utilize a point-of-care bar-coding system, and a failure to employ available bar-coding technology to verify the drug before administration.[13]

The RCA then set out to explore why each of these proximate causes happened, working its way from the sharp end of the error to the underlying system problems that contributed to the error. It found that there was no system for communicating the pain management plan of care for the laboring patient to the nurse responsible for getting the patient ready for an epidural. It also found variable expectations from anesthesia staff regarding patient readiness for an epidural and staff scheduling policies that did not guard against excessive fatigue. It cited the interchangeability of tubing used for epidural and intravenous (IV) solutions. And it found only a 50 percent unit-wide compliance rate with scanning medications using available bar-coding technology.[14]

There is a wonderful side-effect to doing an RCA. And that is that it actually gets people around the table to discuss and explore the deeper issues behind "human error." In healthcare, these people (from different layers in the medical competence hierarchy, from different departments, specializations and professions) would otherwise not have much reason to talk to one another and share their perspectives and thoughts. The dominant logic of an RCA, however, keeps bringing people back to the fixable properties of their sub-systems. The recommendations stemming from the RCA in this case included designing a system to communicate the anesthesia plan of care, defining patient readiness for an epidural, establishing dedicated anesthesia staff for obstetrics, differentiating between epidural and IV medications, designing a quiet zone for preparing medications, establishing maximum work-hour policies for staffing schedules, and remedying issues with scanning problematic containers (such as the translucent IV bags in the case described here) to improve bar- code-scanning compliance rates.[15]

Yet, as John Stoop is fond of saying, there is a difference between explanatory variables and change variables. In other words, there is a difference between those things that can explain why a *particular* event happened, and those things we should focus our attention on to make sure that *similar* things don't happen again. In complex systems, separating these out probably makes great sense. The value (see Chapters 6 and 7) of doing retrospective analyses is limited because of the constant dynamics and unforeseeabilities of complexity: that exact failure, in precisely that sequence, will be very unlikely to recur.

What is more, fixing properties of a system that brought this failure into being may actually reverberate somewhere else, at some other time. This is a standard feature of complexity. Strengthening something in one corner can lead to vulnerabilities elsewhere. This happens because making changes to some components in a complex system leads to an explosion in relationships and changes in relationships with other components (see Chapter 6). For example, establishing a dedicated anesthesia staff for obstetrics might mean that an anesthetic staff crunch occurs elsewhere at particular times. Or that maximum work-hour policies (part of the same recommendations) get into trouble in another ward because of shortages of available competence after the creation of a dedicated staff for obstetrics. The interconnectedness of a complex system makes the fixing of broken parts problematic, or at least acutely interesting. The change to a part, after all, hardly ever stays with that part. It reverberates through all kinds of other parts in ways that are sometimes foreseeable, but often not.

The very incident that eventually led to the criminalization of this nurse started with such an improvement. A couple of years before, the hospital administration had noticed how anesthetists were very unhappy with the working practices at obstetrics. Women in labor could only have their epidural refilled or changed by an anesthetist: midwives and nurses were not allowed to do so. Epidural analgesia used to be given to patients from containers containing 60 cc. These came right from the manufacturer and had a dedicated port for connection to the entry point in the patient's back. The problem was that the containers weren't very large and used to run out in the middle of the night. That meant that anesthetists had to be called down from elsewhere in the hospital, at times when they might just be able to catch some shut-eye. Just to change or renew an epidural.

After a staff survey, a proposal was developed by anesthesia to have the hospital pharmacy itself fill normal 250 cc intravenous bags with epidural analgesia fluid. The pharmacy was happy to do so. The time between epidural changes was more than tripled, and anesthetists became much more content. It may have had positive consequences for the continuity and workflow and even patient safety in other parts of the hospital. In complexity theory, such a small change (or new initial conditions) creates sensitive dependency, however. When new technology was introduced shortly before the death of the 16-year-old, it became obvious that the translucent IV bags were virtually impossible to scan with a barcode scanner (nurses had to find and hold a white piece of paper or cloth behind it so as to bring out the contrast for the scanner to pick up). But a more sinister sensitive dependency was introduced: the interchangeability of the IV bag ports. It was possible to now hook up a bag full of analgesic and let it flow straight into the patient through a normal IV connection.

Most in health care readily embrace the idea that they work in a complex system. The concept of health care as a complex system seems to be widely recognized. Other fields too, are quick to claim the label "complex" to describe their own operations. Yet there is a lingering tendency to reach for simple solutions, for silver bullets, for single-factor explanations. Healthcare in particular is swift to bemoan the "ineptitude" of those defeated by the system's complexity and to celebrate the "strength of character" of those able to bridle it, or to work around it.[16]

If systems really are complex (and see Chapter 6 for more formal definitions), let's start to live with that. We should start to act as if we really understood what that means. Complexity theory, rather than Newtonian reductionism, is where we should look for directions. That is what we should use to consider complex systems that risk drifting into failure. As said, with the introduction of each new part or layer of defense, technology, procedure, or specialization, there is an explosion of new relationships between parts and layers and components that spreads out through the system. Complexity theory explains how accidents emerge from these relationships, even from perfectly "normal" relationships, where nothing (not even a part) is seen as broken. The drive to make systems reliable, then, also makes them very complex – which, paradoxically, can in turn make them less safe. Redundancy or putting in extra barriers, or fixing them does not provide any protection against a system safety threat. In fact, it can help create it, or perpetuate or even heighten the threat. The introduction of a layer of technology (bar-code-scanning) for double-checking a medication order against a patient ID, for example, introduces new forms of work and complexity (the technology doesn't work as advertised or hoped, it takes time and attention away from primary tasks, and it calls for new forms of creativity and resourcefulness).

Newton has been on a retainer for more than three centuries. There is something seductive about going down and in to find the broken part and fix it. We can try to tell professionals to be "more professional," for example, or give them more layers of technology to forestall the sorts of component failures we already know about (only to introduce new error opportunities and pathways to failure). Complexity theory says that if we really want to understand failure in complex systems, that we "go up and out" to explore how things are related to each other and how they are connected to, configured in, and constrained by larger systems of pressures, constraints, and expectations.

We would ask why the nurse in question is at work already again this day after a break of only a few hours (that she spent trying to sleep in an empty hospital bed). We would find that she was filling in an empty slot created by the medical leave of a colleague, on a holiday weekend. Just below, we would find how the subtle but pressing requests to stay for another shift intersects with cultural and personal and deontological features of those we make into our nurses – of those whom we *want* to be our nurses, those who somehow incarnate commitment, dedication, those who are the embodiment of the "care" in healthcare.

We could trace such a situation to various managerial, administrative, political, and budgetary motivations of a hospital, which we could link to insurance mercantilism, the commercialization of disease, the demand for a commodification of health care's prices and products. We would want to find how, since Florence Nightingale, nursing has steadily lost status, reward, and attraction, with ranks that are hard to fill, its traditional provision of succor eroded under the relentless industrialization of care, and its role as patient voice, as patient advocate now hollow, because there is always the next patient. And the next. And, if we have the societal courage, we might inquire after the conditions and collective norms that make it plausible for a 16-year-old girl among us in the community to be pregnant and in need of hospital care to begin with.

If we don't dare to go there and undertake this line of inquiry, then it should be no surprise that the cumulative consequences suddenly emerge one day on the work floor of a busy, understaffed ward in a regional community hospital with a patient screaming in acute, severe pain, demanding that something be done now, *now*. If we tinker only gingerly with the final, marginal technical minutiae at various sharp ends, all of those systemic influences will collect again and again to shape what any other caregiver will see as the most rational course of action – no matter how large the label on the IV bag.

REFERENCES

1   State of Wisconsin, Circuit Court Dane County (Filed 6 November 2006). *Criminal Complaint, Case number 2006 CF 2512*, p. 3.

2   State of Wisconsin, Circuit Court Dane County (Filed 6 November 2006). Ibid., p. 6.

3   Institute for Safe Medication Practices (2006, November 7). *ISMP opposes criminal charges for Wisconsin nurse involved in medication error.* Huntingdon Valley, PA: ISMP.

4   In a previous book, I have tried to suggest ways of thinking about accountability that do not harm safety. See: Dekker, S.W.A. (2007). *Just culture: Balancing accountability and safety.* Aldershot, UK: Ashgate Publishing Co.

5   Capra, F. (1982). *The turning point.* New York: Simon and Schuster, chapter 2.

6   Outram, D. (2005) *The Enlightenment* (Second Edition: New approaches to European history). Cambridge, UK: Cambridge University Press.

7   The full title is *Philosophiae Naturalis Principia Mathematica*. It was published in 1687 and is considered to be among the most influential books in the history of science and the groundwork for most of classical mechanics. Besides considerations of how to conduct scientific experiments, the three laws of motion show up in this book, too.

8   The Latinized form of René Descartes' name was *Renatus Cartesius*, from which the adjective "Cartesian" comes.

9   Vaughan, D. (2005). System effects: On slippery slopes, repeating negative patterns, and learning from mistake? In W.H. Starbuck and M. Farjoun (eds.). *Organization at the limit: Lessons from the Columbia disaster*, pp. 41–59. Malden, MA: Blackwell Publishing.

10   Feynman, R.P. (1988). *'What do you care what other people think?': Further adventures of a curious character.* New York: Norton.

11   Vaughan, D. (2005). Ibid.

12   Leveson, N., Cutcher-Gershenfeld, J., Barrett, B., Brown, A., Carroll, J., Dulac, N., Fraile, L., and Marais, K. (2004). Effectively Addressing NASA's Organizational and Safety Culture: Insights from Systems Safety and Engineering Systems. *Paper presented at the Engineering Systems Division Symposium, MIT*, Cambridge, MA, March 29–31, p. 1.

13   Smetzer, J., Baker, C., Byrne, F., Cohen, M.R. (2010). Shaping systems for better behavioral choices: Lessons learned from a fatal medication error. *The Joint Commission Journal on Quality and Patient Safety*, 36(4), 152–163.

14   Smetzer, J., Baker, C., Byrne, F., Cohen, M.R. (2010). Ibid.

15   Smetzer, J., Baker, C., Byrne, F., Cohen, M.R. (2010). Ibid.

16   See: Gawande A. (2002). *Complications: A Surgeon's Notes on an Imperfect Science*. New York City: Picador; and Pellegrino, E.D. (2004). Prevention of medical error: Where professional and organizational ethics meet. In Sharpe, V.A. (ed.). *Accountability: Patient Safety and Policy Reform*. Washington, DC: Georgetown University Press, pp. 83–98.

4

THE SEARCH FOR THE BROKEN COMPONENT

If there is one idea that Newton and Descartes have taught us when things go wrong, it is to look for the broken component. Nothing characterizes the Newtonian–Cartesian worldview as much as its emphasis on analytic reduction, or decomposition. The story in Chapter 2, of course, is about a broken part – the jackscrew/acme nut on an MD-80 airliner. Though perhaps it is more correct to say that the story starts with a broken part. In itself the part isn't that interesting. If we want to understand *why* it ended up broken, analytic reduction doesn't get us very far. Instead, we need to go up and out, rather than down and in. We have to begin to probe the hugely intertwined webs of relationships that spring out and away from the broken part, into the organizational, the institutional, the social. Yet often our quest to understand why parts are broken simply leads us to other broken parts. The decompositional logic is almost everywhere.

Consider the practice, very common in all kinds of accident investigations, to examine the 24-hour and 72-hour histories of the persons involved at the sharp end – the train driver, the pilots, the first mate. Hardly ever do these examinations reveal anything interesting. Sure, the pilot may have been playing tennis with her husband the day before, got 7 hours of sleep before getting up to report to the airport. The train driver had a 10-hour day prior to the day of the accident. So? Well, this is to rule out anything to do with the operators, investigators have told me. Fair enough.

But this seemingly very innocent practice of excluding possible causes serves to reconfirm the image of an investigation as a search for the broken component in the rubble. It legitimates as well as reproduces a kind of naïve Newtonian physics about how systems work and fail. How do we find broken components? The 24- and 72-hour histories suggest that we first start with the people at the sharp end. We rummage through their lives, we probe more and more hours backwards into their professional and private doings, hoping for the "Aha!" part to pop up. A pill bottle on the night stand. A fight with the spouse. A lousy hotel room on an overnight with too much noise and too little sleep. That last beer leaking over into the 8-hour window from bottle to throttle.

If this search doesn't turn up anything; if we fail to find the broken component there, we redirect our attention to the next system, or the next level. We may end up by looking at the Safety Management System of the organization, a fashionable thing to do nowadays. And we may find that it is broken in various places. Reporting doesn't work as well, for example, or the quality management of a training program is deficient in how it tracks people's progress along various levels of documentation and bureaucratic accountability. The resulting investigation, a catalogue of broken or bent components, is all too common. The "findings" section of accident reports are often precisely such an engineering catalogue, listing the components that were found broken, and the components that were found not broken.

BROKEN COMPONENTS AFTER A HAILSTORM

After its investigation of an incident in which an airliner flew into a hailstorm and sustained severe damage, the report came up with the following in its "findings" section (among others):

1. The pilots' licenses and ratings were valid.
OK, component not broken.

2. The aircraft had a valid certificate of airworthiness.
OK, component not broken.

3. The pilots reported for duty well before the required time.
OK, component not broken.

4. The pilots did not study or take with them the Nordic Severe Weather Chart for the time of their flight. Moreover, they did not study the Severe Weather Chart prepared by the World Area Forecast Centre.
Not OK – two broken components found.

5. The Aeronautical Weather Services was preparing a SIGMET (a forecast for significant weather) but it was published too late with regard to the development of weather conditions.
Not OK – one broken component found.

6. Engine anti-ice and ignition systems were used as instructed.
OK, component not broken.
7. The galleys in the cabin had been secured for landing.
OK, component not broken.

8. The pilot flying did not disengage the autothrottle when the aircraft flew into turbulence, as recommended by the Aircraft Operating Manual.
Not OK, one broken component found.

9. The co-pilot tried to switch the weather radar on, but no radar image was shown.
Not OK – one broken component found.

10. After the windshields had been cracked, the pilot flying turned the windshield heating switch off. This action is not included in the malfunction check list flight deck window cracks in flight. The flight crew did not take the actions mentioned in the checklist.
Not OK – one broken component found.

11. The cabin crew paid attention to the abnormal, rattling sound of the engines. However, they did not inform the cockpit crew of their observation during the flight.
Not OK – one broken component found.

And so it goes on. A catalogue of components, both broken and not broken.

An accident report that lists findings as its major product invokes comparison with the way in which we used to train doctors (and still do, in many places). Give the students a bunch of components (both broken and not broken), divvy them up across classes ranging from anatomy to physiology to neurology, and then let the students themselves piece it all together; let them figure out how all these components work together to account for the functioning or malfunctioning of a human body. In case of a typical accident or incident report, the writers make a similar leap of faith. The reader is trusted to make a meaningful connection between a list of broken and not-broken components and a coherent narrative of the event. This connection only works in a Newtonian–Cartesian world, where there is a straightforward relationship between the way in which components work or fail, and how the system works or fails as a whole.

The perpetual search for broken components is in part a heritage of the engineering origins of accident investigations. An engineering understanding of the world, at least if it is driven by Newtonian–Cartesian logic, is to break the large thing (the accident or incident) apart so as to reveal its inner components. Ferret about among those components, and find the ones that are broken.

Note the really interesting (and, as you will see later, really problematic) assumption that there is a direct relationship between the broken component(s) inside the system, and the behavior of the system as a whole. In the example above, the supposed mishandling of the encounter with hail by the flight crew, and the resulting damage and danger to the aircraft and its occupants, can be reduced to the pilots not conducting some of the procedures by the book, and making other procedures up where the book offered none. The whole, in other words, can be explained by reference to the behavior of the parts. The functioning or non-functioning of the whole can be understood through the functioning or non-functioning of constituent components. In the Newtonian–Cartesian world, their relationship is unproblematic, direct, and in no need of further proof or synthetic work. This, in effect, is where many investigations close the book. Broken component(s) identified, fixes suggested. Finished.

On the other hand, we could also consider our system as something that is alive, as something that has all kinds of processes running through it at a variety of levels, which connect the various components in many complex ways. In that case, pulling the system apart and lifting components out of it to examine them for their individual performance basically kills the phenomenon of interest. Broken components may not really explain a broken system. But we are getting ahead of ourselves. For now, let's try to understand how the hunt for broken components is something that we equate with good analysis, with finding the cause, with knowing what went wrong, with knowing what to fix. So what if we can't find a broken component?

In the summer of 1996, Trans World Airlines flight 800 from New York to Paris, a Boeing 747–100 with 230 people onboard, exploded suddenly when climbing out over the sea alongside Long Island. It was only 11 minutes after takeoff. The huge amount of kerosene for its Atlantic crossing vaporized and ignited, creating a fireball that was seen along the coastline of Long Island. The fragmented aircraft disappeared into the ocean, forming an expanding bubble of debris so pulverized that a weather radar interpreted it as a rain cloud.

What followed was one of the most expensive and most vexing accident investigations to date. Yet, despite its length and the reconstruction of the entire wreckage in a hangar close by, it failed to yield the "eureka part," that single bit of wreckage, or equipment, that contained a critical clue, a clue that could point to the broken component. It gave rise to a large number of conspiracy theories. A light streaking towards the jet was seen by some witnesses just prior to the explosion, metal in one place seemed bent inwards and then outwards, pointing to a hole created by an external projectile, and microscopic traces of PETN (Pentaerythritol Tetranitrate), a compound used in plastic explosives, were found on a piece of flooring from the passenger compartment. A conspiracy involving the U.S. Navy, a test missile, the FBI and a whole host of other agencies, would be one reasonable candidate cause. Later, however, the FBI stated that these traces were consistent with explosives allegedly spilled during a training exercise aboard the aircraft a few weeks before the crash.

After more than four years, the National Transportation Safety Board (NTSB) concluded that the probable cause of the accident was an explosion of the center wing fuel tank, resulting from ignition of the flammable fuel/air mixture in the tank. The source of ignition energy for the explosion could not be determined with certainty, but, of the sources evaluated by the investigation, the most likely was a short-circuit outside of the tank that allowed excessive voltage to enter it through electrical wiring associated with the fuel quantity indication system. Researchers from across the U.S.A., dissatisfied with the many anomalies and unexplained pieces of evidence in the official investigation, have gathered in a group called the "Flight 800 Independent Researchers Organization," (or FIRO) dedicated to assisting official investigators determine "the cause" of this yet unresolved disaster.

Interestingly, the chairman of the NTSB at the time, Jim Hall, raised the specter of his agency not being able to find the eureka part, which would challenge its entire reputation: "What you're dealing with here is much more than an aviation accident because of the profile of the crash. What you have at stake here is the credibility of this agency and the credibility of the government to run an investigation," he said.

In other words, the credibility of the U.S. government to run an investigation could hinge on its ability to return with a eureka part in its hands. A failure to produce a eureka part, conversely, would mean that the government's approach, methods, techniques and expertise were insufficient or inadequate. The eureka part was out there, but it would take, as FIRO aimed to do, a different kind of investigation, less political, less cowed by other agencies, and following up on other clues, ironing out inconsistencies and ambiguities that had been left in place.

In a sense, the credibility of the government to run an investigation hinged on its ability to find the cause for the effect, to find the "eureka part," to locate the part that would have everybody in the investigation declare that they'd found the broken component, the trigger, the original culprit, that could carry the explanatory load of the loss of the Boeing 747. But for this crash, the so-called "eureka part" was never found. The FIRO (Flight 800 Independent Researchers Organization) position represented the culmination of dissatisfaction with this inability. Theirs represented a very Newtonian commitment, just like the NTSB's. For FIRO, the eureka part was out there; all it would take was a better method, or a different method. The idea that accidents could be produced without any part failing, or without anything external interfering, but rather as a result of normally functioning components interacting in unforeseen ways,[1] was not up for debate here. It would have fallen outside the ruling paradigm: an effect must have a cause, and that cause should be a broken component, somewhere, somehow, that can be found. Even if it took belief in a conspiracy to come up with one.

BROKEN COMPONENTS TO EXPLAIN A BROKEN SYSTEM

Componential explanations that condense accounts of failure down to the failure of a part still reign supreme. And the human is often that part. A recent analysis by Richard Holden showed that between 1999 and 2006, 96 percent of investigated U.S. aviation accidents were attributed in large part to the flight crew. In 81 percent, people were the *sole* reported cause. The language used in these analyses has judgmental or even moralistic overtones too. "Crew failure" or a similar term appears in 74 percent of probable causes and the remaining cases contain language such as "inadequate planning, judgment and airmanship," "inexperience" and "unnecessary and excessive ... inputs." "Violation" of written guidance was implicated as cause or contributing factor in a third of all cases.[2] Single-factor, judgmental explanations for complex system failures are not unique to aviation – they are prevalent in fields ranging from medicine, shipping, military operations, to road traffic.[3]

The problem of safety analysis reverting to condensed and individual/ componential explanations rather than diffuse and system-level ones was one driver behind the fields of human factors and system safety. A charter of these fields has been to take safety scientists and practitioners behind the label "human error," to more complex stories of how normal system factors contribute to things going wrong (including factors associated with new technology and with the organization and its multiple interacting goals). But common discourse about failure in complex systems remains tethered to language such as "chain-of-events," "human error" and questions such as "what was *the* cause?" and "*who* was to blame?"

NEWTON AND THE SIMPLICITY OF FAILURE

The search for broken or underperforming components that can carry the explanatory and moral load of accidents in complex systems has deep roots in what Western society finds rational, logical and just. As shown in the previous chapter, throughout the past three and a half centuries, the West has equated scientific thinking with a Newtonian–Cartesian worldview which prizes decomposition, linearity, and the pursuit of complete knowledge. Let's look at how this has given rise to what I call a Newtonian ethic of failure – which makes particular assumptions about the relationship between cause and effect, foreseeability of harm, time-reversibility and the ability to come up with the "true story" (that is, "the" cause) of a particular event. These assumptions animate much of our safety work today, including accident investigations, societal expectations, managerial mandates, technical tools and artifacts, and judicial responses to system failures.

The logic behind Newtonian science is easy to formulate, although its implications for how we think about the ethics of failure are subtle, as well as pervasive. In this section we review aspects of the Newtonian–Cartesian worldview that influence how we understand (and consider the ethics around) failure in complex systems, even today. Until the early twentieth century, classical mechanics, as formulated by Newton and further developed by Laplace and others, was seen as the foundation for science as a whole. It was expected that observations made by other sciences would sooner or later be reduced to the laws of mechanics. Although that never happened, other disciplines, including psychology and law, did adopt a reductionist, mechanistic or Newtonian methodology and worldview. This influence was so great that most people still implicitly equate "scientific thinking" with "Newtonian thinking." The mechanistic paradigm is compelling in its simplicity, coherence and apparent completeness. It was not only successful in scientific applications, but also largely consistent with intuition and common sense.

REDUCTIONISM AND THE EUREKA PART

As said, the best known principle of Newtonian science, formulated well before Newton by the philosopher-scientist Descartes, is that of analysis or reductionism. To understand any complex phenomenon, you need to take it apart, that is, reduce it to its individual components. This is recursive: if constituent components are still complex, you need to take your analysis a step further, and look at their components. In other words, the functioning or non-functioning of the whole can be explained by the functioning or non-functioning of constituent components. Attempts to understand the failure of a complex system in terms of failures or breakages of individual components in it – whether those components are human or machine – are very common.

Linear, componential thinking permeates the investigation of accidents and organizational failures as well as reliability engineering methods. The defenses-in-depth metaphor, popularized as the "Swiss Cheese Model" and used in event classification schemes relies on the componential, linear parsing-up of a system, so as to locate the layer or part that was broken. The analytic recursion in these methods ends up in categories such as "unsafe supervision" or "poor managerial decision-making." Indeed, in technically increasingly reliable systems, the "eureka part" has become more and more the human and there is a cottage industry of methods that try to locate and classify which errors by which people lay behind a particular problem. Most or all implicitly presume a linear relationship between the supposed error and the parts or processes that were broken or otherwise affected.

Our understanding of the psychological sources of failure is subject to reductive Newtonian recursion as well. In cases where the component failure of "human error" remains incomprehensible, we take "human error" apart too. Methods that subdivide human error up into further component categories, such as perceptual failure, attention failure, memory failure or inaction are now in use in air traffic control and similar linear, reductionist understandings of human error dominate the field of human factors. The classically mechanistic idea of psychology that forms the theoretical bedrock for such reductionist thinking of course predates human factors. Analytic reduction sponsors an atomistic view of complex psychological phenomena: understanding them comes from revealing the functioning or breakdown of their constituent components.

Even sociological or cultural phenomena are often explained using a reductionist approach. Much safety culture work, for example, breaks "culture" down into what individual employees experience or believe (mostly through attitude surveys about, for example, work pressure, management behaviors in relation to safety, risk perceptions), as well as the presence or absence of particular components of, for example, a safety management system, safety investments, worker qualifications or other safety arrangements. It measures those, adds them up and gets a number for a safety culture. Together, these components are assumed to constitute a "culture," yet it is never made clear how the parts become the whole. Such approaches meet with skepticism from those who see culture as something much more complex and incompressible:

> Culture cannot be managed; it emerges. Leaders don't create cultures; members of the culture do. Culture is an expression of people's deepest needs, a means of endowing their experiences with meaning. Even if culture in this sense could be managed, it shouldn't be (…) it is naive and perhaps unethical to speak of managing culture.[4]

This is not to say that measuring these things is not meaningful. It may well identify interesting areas for safety intervention. But to say that it measures "culture" is making a particular set of Newtonian assumptions about the relationship between part and whole, between individual and society. This relationship is called into question by anthropologists like Hutchins, who argue that the regularities that are seen as characteristic for a culture (the whole) cannot be easily found inside the members (the parts):

> A given group of individuals may enact different distributed cognitive processes depending on institutional arrangements. Observed patterns of behavior emerge from the interactions of the internal processes with structures and processes that are present in the environment for action. This means that the regularities that are often identified as being characteristic of a culture may not be entirely "inside" the individual members of the society in question and may not generalize across activity settings. From the contextual point of view, the term culture can be read as a shorthand label for an emergent uneven distribution of a variety of material, social, and behavioral patterns that result from a universal human process.[5]

The atomistic view of culture or community or society sees the relationship between parts and whole as unproblematic. The parts simply add up to make the whole. If parts are imperfect or even missing (for example, if particular components of a safety management system are incomplete, or access to the boss with safety concerns is impossible), then this will add up to a culture in a straightforward way, to a measurable safety culture (or lack thereof). This logic (of a simple, additive relationship between parts and the whole) is so pervasive that Margaret Thatcher even denied the existence of community or society altogether in an interview in 1987: " … there is no such thing as society. There are individual men and women, and there are families." From that extreme point of view, there are only atoms, parts, components. And somehow, they make up a whole. But for that point of view, the whole isn't the point. It's the parts and the easily added numbers they represent.

It is interesting to note that such a position is quite consistent with Western, and often Protestant societies. Through the Enlightenment, and the intellectual tradition since the Scientific Revolution, it has seemed self-evident to evaluate ourselves as individuals, bordered by the limits of our minds and bodies, and that we get assessed in terms of our own personal achievements. From the renaissance onwards, the individual became a central focus, fueled in part by Descartes' psychology that created self-contained individuals. The rugged individualism developed on the back of mass European immigration into North America in the late nineteenth and early 20th centuries accelerated the image of independent, free heroes accomplishing greatness against all odds, and anti-heroes responsible for disproportionate evildoing. The notion that it takes teamwork, or an entire organization, or an entire industry (think of Alaska 261) to break a system is just too eccentric relative to this cultural prejudice.

The philosophy of Newtonian science is one of simplicity: the complexity of the world is only apparent and to deal with it we need to analyse phenomena into their basic components. The way in which legal reasoning in the wake of accidents separates out one or a few actions (or inactions) on the part of individual people follows such reductive logic. For example, the Swedish Supreme Court ruled that if one nurse had more carefully double-checked a particular medication order before preparing it (mistakenly at 10 times the intended dose) a three-month old baby would not have died. Such condensed, highly focused accounts that converge on one (in)action by one person (the "eureka part") give componential models of failure a societal legitimacy that keeps reproducing and instantiating Newtonian physics.

CAUSES FOR EFFECTS CAN BE FOUND

In the Newtonian vision of the world, all that happens has a definitive cause and a definitive effect. In fact, there is a symmetry between cause and effect (they are equal but opposite). The determination of the "cause" or "causes" is, of course, seen as the most important function of accident investigation today, and assumes that physical effects (a crashed airliner, a dead patient) can be traced back to physical causes (or a chain of causes-effects). The notion that effects cannot occur without causes makes it into legal reasoning in the wake of accidents too. For example, "to raise a question of negligence, harm must be caused by the negligent action."[6] It is assumed that a causal relationship (that negligent action caused harm) is indeed demonstrable, provable beyond reasonable doubt.

The Newtonian view of the world that holds all this up is materialistic: all phenomena, whether physical, psychological or social, can be reduced to matter, that is, to the movement of physical components inside three-dimensional Euclidean space. The only property that distinguishes particles is where they are in that space. Change, evolution, and indeed accidents, can be reduced to the geometrical arrangement (or misalignment) of fundamentally equivalent pieces of matter, whose interactive movements are governed exhaustively by linear laws of motion, of cause and effect. A visible effect (for example, a baby dead of lidocaine poisoning) cannot occur without a cause (a nurse blending too much of the drug). The Newtonian assumption of proportionality between cause and effect can in fact make us believe that really bad effects (the dead baby) have really bad causes (a hugely negligent action by an incompetent nurse). The worse the outcome, the more "negligent" its preceding actions are thought to have been. In road traffic, talking on a cell phone is not considered illegal by many, until it leads to a (fatal) accident. It is the effect that makes the cause bad.

The Newtonian model has been so pervasive and coincident with "scientific" thinking, that if analytic reduction to determinate cause-effect relationships (and their material basis) cannot be achieved, then either the method or the phenomenon isn't considered worthy of the label "science." This problem of scientific self-confidence has plagued the social sciences since their inception, inspiring not only Durkheim to view the social order in terms of an essentialist naïve Newtonian physics, but also for example to have Freud aim "to furnish a psychology that shall be a natural science: that is, to represent psychical processes as quantitatively determinate states of specifiable material particles, thus making those processes perspicuous and free from contradiction."[7] Behaviorists like Watson reduced psychological functioning to mechanistic cause-effect relationships in a similar attempt to protect social science from accusations of being unscientific.[8]

THE FORESEEABILITY OF HARM

According to Newton's image of the universe, the future of any part of it can be predicted with absolute certainty if its state at any time was known in all details. With enough knowledge of the initial conditions of the particles and the laws that govern their motion, all subsequent events can be foreseen. In other words, if somebody can be shown to have known (or should have known) the initial positions and velocities of the components constituting a system, as well as the forces acting on those components (which in turn are determined by the positions of these and other particles), then this person could, in principle, have predicted the further evolution of the system with complete certainty and accuracy. A system that combines the physiology of a three-month old baby with the chemical particles diethylamino-dimethylphenylacetamide that constitute lidocaine will follow such lawful evolution, where a therapeutic dose is less than 6 mg lidocaine per gram serum, and a dose almost 10 times that much will kill the baby.

If such knowledge is in principle attainable, then the harm that may occur if particles are lined up wrongly is foreseeable, too. Where people have a duty of care (like nurses and other healthcare workers do) to apply such knowledge in the prediction of the effects of their interventions, it is consistent with the Newtonian model to ask how they failed to foresee the effects. Did they not know the laws governing their part of the universe (that is, were they incompetent, unknowledgeable)? Were they not conscientious or assiduous in plotting out the possible effects of their actions? Indeed, legal rationality in the determination of negligence follows this feature of the Newtonian model almost to the letter: "Where there is a duty to exercise care, reasonable care must be taken to avoid acts or omissions which can reasonably be foreseen to be likely to cause harm. If, as a result of a failure to act in this reasonably skillful way, harm is caused, the person whose action caused the harm, is negligent."[9]

In other words, people can be construed as negligent if the person did not avoid actions that could be foreseen to lead to effects — effects that would have been predictable and thereby avoidable if the person had sunk more effort into understanding the starting conditions and the laws governing the subsequent motions of the elements in that Newtonian sub-universe. Most road traffic legislation is founded on this Newtonian commitment to foreseeability too. For example, a road traffic law in a typical Western country might read how a motorist should adjust speed so as to be able stop the vehicle before any hinder that might be foreseeable, and remain aware of the circumstances that could influence such selection of speed.[10] Both the foreseeability of all possible hinders and the awareness of circumstances (initial conditions) as critical for determining speed are steeped in Newtonian epistemology. Both are also heavily subject to outcome bias: if an accident suggests that a hinder or particular circumstance was not foreseen, then speed was surely too high. The system's user, as a consequence, is always wrong. And any search for the cause of failure will therefore always turn up a broken part.

TIME-REVERSIBILITY

The trajectory of a Newtonian system is not only determined towards the future, but also towards the past. Given its present state, we can in principle reverse the evolution to reconstruct any earlier state that it has gone through. The Newtonian universe, in other words, is time-reversible. Because the movement of, and the resulting interactions between, its constituent components are governed by deterministic laws of cause and effect, it does not matter what direction in time such movements and interactions are plotted. Such assumptions, for example, give accident and forensic investigators the confidence that an event sequence can be reconstructed by starting with the outcome and then tracing its causal chain back into time. The notion of *re*construction reaffirms and instantiates Newtonian physics: our knowledge about past events is nothing original or creative or new, but merely the result of uncovering a pre-existing order. The only thing between us and a good reconstruction are the limits on the accuracy of our representation of what happened. We then assume that this accuracy can be improved by "better" methods of investigation, for example.

COMPLETENESS OF KNOWLEDGE

The traditional Western belief in science is that its facts have an independent existence outside of people's minds: they are naturally occurring phenomena "out there," in the world. The more facts a scientist or analyst or investigator collects, the more it leads, inevitably, to more, or better, science: a better representation of "what happened." The belief is that people create representations or models of the "real" out there, models or representations that mimic or map this reality. Knowledge is basically that representation. When these copies, or facsimiles, do not match "reality," it is due to limitations of perception, rationality, or cognitive resources, or, particularly for investigators or researchers, due to limitations to methods of observation. More, or more refined methods and more data collection, can compensate for such limitations.

Newton argued that the laws of the world are discoverable and ultimately completely knowable. God created the natural order (though kept the rulebook hidden from man; instead God gave man intelligence to go figure it out for himself) and it was the task of science to discover this hidden order underneath the apparent disorder. The Newtonian view is based on the reflection-correspondence view of knowledge: our knowledge is an (imperfect) mirror of the particular arrangements of matter outside of us.[11] The task of investigations, or science, is to make the mapping (or correspondence) between the external, material objects and the internal, cognitive representation (for example, language, or some mental model) as accurate as possible. The starting point is observation, where information about external phenomena is collected and registered (for example, the gathering of "facts" in an accident investigation), and then gradually completing the internal picture that is taking shape. In the limit, this can lead to a perfect, objective representation of the world outside.[12] The world is already there, pre-formed and pre-existing. All we need to do is uncover and then describe the order that the world already possesses, that it was given by its creator.

The Newtonian position can be recognized in what society generally sees as high science today. The leader of the International Human Genome Sequencing Consortium and currently director of the U.S. National Institutes of Health recently wrote a book entitled *The Language of God*.[13] For him, the sequencing of the human genome (our genetic information, stored on 23 chromosomes and occupied by more than three billion DNA base pairs) was ultimately a discovery of God's language: God's code in which the essence of humanity had been written. The language, the order, had been there for thousands of millennia already. But at the closing of the second millennium CE, science had finally arrived at a method to lay it bare. It is a position that inspires awe and respect: people can marvel in the amazing creation.

But the human genome project revealed that the supposedly pre-existing order of the building blocks of life (DNA) severely underdetermines how the complex system looks, or works, or fails. Only about 1 percent of the 3 billion base pairs actually code for human features that we know about. Humans have fewer than double the number genes of very basic organisms such as fruit flies and roundworms, and have more than 99 percent genes in common with chimpanzees. Genetically, we are even closer to Neanderthals. This is typical for complex, as opposed to Newtonian, systems. Complexity implies an ultimately intractable relationship between the parts and the whole.

What is more, the language that the Human Genome Sequencing Consortium uncovered was actually hardly orderly, and hardly entirely pre-existing. Human cells make extensive use of alternative splicing by which a single gene can code for multiple proteins, which already explodes any straightforward relationship between parts and whole. There also seem to be nonrandom patterns of gene density along chromosomes which are not well understood, and regions of coding and non-coding DNA. Then there is gene-switching: the ability to turn genes on or off through the organism's interaction with its changing environment.

The human genome, in other words, can at best be described as a set of hardware, which in turn can run all kinds of versions of software and thousands of different and partially overlapping and even contradictory programs at the same time. Complex systems are open systems, and so is the human genome. It is this openness to the environment and the system's ability to recognize, adapt, change, and respond that renders any project to describe the pre-existing order quite hopeless. When you think you've described the system once, you'll find that it will have morphed away from your description before you're even done. And even any temporary or tentative description of the arrangement of parts spectacularly underdetermines what can be observed at the system level.

Founding sociologist Emile Durkheim took the same Newtonian position for social science in the nineteenth century. Underneath a seemingly disordered, chaotic appearance of the social world, he argued, there is a social order governed by discoverable laws (obligations and constraints) and categories of human organization (institutions). As with the theological reading of the human genome project a century-and-a-half later, Durkheim wanted to pursue the essential properties of social systems: those features that are enduring, unchanging, and that can be discovered and described independent on who does the discovering or describing, from what angle or perspective, and at what point in time. Such "essentialism" is typical for Newtonian science: the idea that behind our initial befuddlement and confusion of the world as it meets us, lies an unchanging, pre-existing order of hard facts that we can lay bare in due time.

The consequence of this Newtonian position for understanding failure is that there can be only one true story of what happened. For Newton and Descartes, the "true" story is the one in which there is no more gap between external events and their internal representation. Those equipped with better methods, and particularly those who enjoy greater "objectivity," (that is, those who, without any bias that distorts their perception of the world, will consider *all* the facts) are better poised to achieve such a true story. Formal, government-sponsored accident investigations enjoy this aura of objectivity and truth – if not in the substance of the story they produce, then at least in the institutional arrangements surrounding its production. First, their supposed objectivity is deliberately engineered into the investigation as a sum of subjectivities. All interested parties (for example, vendors, the industry, the operator, the legal system, unions, professional associations) can officially contribute (though some voices are easily silenced or sidelined or ignored). Second, those other parties often wait until a formal report is produced before publicly taking either position or action, legitimating the accident investigation as arbitrator between fact and fiction, between truth and lie. It supplies the story of "what really happened." Without first getting that "true" story, no other party can credibly move forward.

A NEWTONIAN ETHIC OF FAILURE

Together, taken-for-granted assumptions about decomposition, cause-effect symmetry, foreseeability of harm, time reversibility, and completeness of knowledge give rise to a Newtonian ethic in the wake of failure. It can be summed up as follows:

○   To understand a failure of a system, we need to search for the failure or malfunctioning of one or more of its components. The relationship between component behavior and system behavior is analytically non-problematic.

○   Causes for effects can always be found, because there are no effects without causes. In fact, the larger the effect, the larger (for example, the more egregious) the cause must have been.

○   If they put in more effort, people can better foresee outcomes. After all, they would better understand the starting conditions and are already supposed to know the laws by which the system behaves (otherwise they wouldn't be allowed to work in it). With those two in hand, all future system states can be predicted. If they can be predicted, then harmful states can be foreseen and should be avoided.

○   An event sequence can be reconstructed by starting with the outcome and tracing its causal chain back into time. Knowledge thus produced about past events is the result of uncovering a pre-existing order.

○   There can be only one true story of what happened. Not just because there is only one pre-existing order to be discovered, but also because knowledge (or the story) is the mental representation or mirror of that order. The *truest* story is the one in which the gap between external events and internal representation is the smallest. The *true* story is the one in which there is no gap.

As in many other fields of inquiry, these assumptions remain largely transparent and closed to inquiry in safety work precisely because they are so self-evident and common-sensical. The way they get retained and reproduced is perhaps akin to what the sociologist Althusser called "interpellation."[14] People involved in safety work are expected to explain themselves in terms of the dominant assumptions; they will make sense of events using those assumptions; they will then reproduce the existing order in their words and actions. Organizational, institutional and technological arrangements surrounding their work don't leave plausible alternatives (in fact, they implicitly silence them). For instance, investigators are mandated to find the probable cause(s) and turn out enumerations of broken components as their findings. Technological-analytical support (incident databases, error analysis tools) emphasizes linear arrangements and the identification of malfunctioning components. Also, organizations and those held accountable for failure inside them, need something to "fix," which further valorizes condensed accounts, focuses on localization of a few single problems and re-affirms a pre-occupation with components. If these processes fail to satisfy societal accountability requirements, then courts deem certain practitioners criminal, lifting uniquely bad components out of the system.

Newtonian hegemony, then, is maintained not by imposition but by interpellation, by the confluences of shared relationships, shared discourses, institutions, and knowledge. Foucault called the practices that produce knowledge and keep knowledge in circulation an episteme: a set of rules and conceptual tools for what counts as factual. Such practices are exclusionary. They function in part to establish distinctions between those statements that will be considered true and those that will be considered false. Or factual rather than speculative. Or just rather than unjust.[15] The true statement will be circulated through society, reproduced in accident reports, for example, and in books and lectures about accidents. These true statements will underpin what is taken to be common-sensical knowledge in a society – that is, the Newtonian physical order. The false statement will quickly fade from view as it not only contradicts common sense, but also because it is not authorized by people legitimated and trusted by society to furnish the truth.

A naïve socio-technical Newtonian physics is thus continuously read into events that could yield much more complexly patterned interpretations. Newtonian assumptions not only support but also reproduce their own ideas, values, sentiments, images, and symbols about failure, its origin and its appropriate ethical consequences. Collectively, they assert that action in the world can be described as a set of casual laws, with time reversibility, symmetry between cause and effect, and a preservation of the total amount of energy in the system. The only limiting points of such an analysis are met when laws are not sufficiently rigorous or exhaustive, but this merely represents a problem of further methodological refinement in the pursuit of greater epistemological rigor and more empirical data produced because of it.

On April 27, 1900, Lord Kelvin, the eminent physicist and president of Great Britain's Royal Society, confidently told his members that Newton's approach had been successfully extended to embrace all of physics, including both heat and light. In fact, the turn of the century was accompanied by an optimism about Newton's theories of particles and motion, which had been confirmed by generations of scientists in fields ranging from chemistry to biology to sociology.[16] Newton appeared to be able to explain every phenomenon in the universe. Everything was, in principle, explicable, knowable. The great order of the creation would be uncloaked. The outbreak of colossal epidemics had been tamed, the industrial revolution was moving into overdrive. People's infatuation with science and technology gave them supreme confidence in their ability to manipulate and control the world around them.

Today, we have inherited a Newtonian commitment that all but excludes even a common awareness of alternatives. It is not that more complex readings would be "truer" in the sense of corresponding more closely to some objective state of affairs. But they could hold greater or at least a different potential for safety improvement, and could help people reconsider what is useful and ethical in the aftermath of failure. The next chapter studies the scientific literature on safety and accidents for how it has attempted to theorize drift. As we will see, Newtonian assumptions keep cropping up, even in narratives that try to take history and complexity seriously.

REFERENCES

1   Leveson, N., Cutcher-Gershenfeld, J., Barrett, B., Brown, A., Carroll, J., Dulac, N., Fraile, L., and Marais, K. (2004). Effectively Addressing NASA's Organizational and Safety Culture: Insights from Systems Safety and Engineering Systems. *Paper presented at the Engineering Systems Division Symposium, MIT*, Cambridge, MA March 29–31.

2   Holden, R. (2009). People or Systems? To blame is human. The fix is to engineer. *Professional Safety*, 12, 34–41.

3   Catino, M. (2008). A review of literature: Individual blame vs. organizational function logics in accident analysis. *Journal of Contingencies and Crisis Management*, 16(1), 53–62.

4   Martin, J. (1985). Can organizational culture be managed? In Frost, P.J. (ed.), *Organizational Culture*. Thousand Oaks, CA: Sage Publications, p. 95.

5   Hutchins, E., Holder, B., and Pérez, R.A. (2002). *Culture and flight deck operations (Research Agreement 22–5003)*. San Diego: University of California, UCSD, p. 3.

6   GAIN (2004). *Roadmap to a just culture: Enhancing the safety environment*. Global Aviation Information Network (Group E: Flight Ops/ATC Ops Safety Information Sharing Working Group).

7   Freud, S. (1950). Project for a scientific psychology. In *The standard edition of the complete psychological works of Sigmund Freud, vol. I*. London: Hogarth Press, p. 295.

8   Dekker, S.W.A. (2005). *Ten questions about human error: A new view of human factors and system safety*. Mahwah, NJ: Lawrence Erlbaum Associates.

9   GAIN (2004). Ibid., p. 6.

10 This comes from the Swedish *Trafikförordning* 1996:1276, kap.3, §14 and 15. See also Tingvall, C. and Lie, A. (2010). The concept of responsibility in road traffic (Ansvarsbegreppet i vägtrafiken). Paper presented at *Transportforum*, Linköping, Sweden, 13–14 January.

11 Heylighen, F. (1989). Causality as distinction conservation: A theory of predictability, reversibility and time order. *Cybernetics and Systems*, 20, 361–84.

12 Heylighen, F., Cilliers, P., and Gershenson, C. (2005). *Complexity and philosophy*. Vrije Universiteit Brussel: Evolution, Complexity and Cognition.

13 Collins, F.S. (2006). *The language of God*. New York: Free Press.

14 Althusser, L. (1984). *Essays on ideology*. London: Verso.

15 Foucault, M. (1980). Truth and power. In C. Gordon (ed.), *Power/Knowledge*, pp. 80–105. Brighton: Harvester.

16 Peat, F.D. (2002). *From certainty to uncertainty. The story of science and ideas in the twentieth century*. Washington, DC: Joseph Henry Press.

5

THEORIZING DRIFT

MAN-MADE DISASTERS

In 1966, a portion of a coal mine tip (unusable material dug up in the process of mining coal) on a mountainside near Aberfan, South Wales, slid down into the village and engulfed its school. 144 people were killed, including 116 children. The post-disaster inquiry waded into a morass of commissions and bodies and agencies and parties responsible (or not) for the various aspects of running a coal mine, including the National Coal Board, the National Union of Mineworkers, the local Borough Council, the local Planning Committee, the Borough's engineering office, the Commission on Safety in Mines, Her Majesty's Inspector of Mines and Quarries and a local Member of Parliament. Collectively, there was a belief that tips posed no danger; that mining was dangerous for other reasons (which were generally believed to be well-controlled through these multifarious administrative and regulatory arrangements). While danger in the tip grew, the belief that everything was okay remained in place. Until the tip slid into the village.

A few years later, at the University of Exeter, a researcher called Barry Turner found that the inquiry reports of this and other disasters contained a wealth of data about the administrative shortcomings and failures that precipitated those accidents. Yet there was no coherent framework, no *theory*, that could do something sensible with all this data. His book *Man-Made Disasters* changed that. It came out in 1978.[1] It had begun as a doctoral dissertation, based on a qualitative analysis of 84 British accident inquiry reports (which bore the title *Failure of Foresight* – considered as subtitle of the subsequent book).

Man-Made Disasters was an opening shot in theorizing drift. This chapter will run through ideas in the literature that have tried to theorize drift one way or other. In these ideas, drift is virtually always associated with a shift in norms, a shift in what is considered acceptable. It has to do with a gradual, virtually unnoticed erosion of safety constraints, with the complexity and lack of transparency of organizations, with limits on human and social rationality and with the pre-rational, unacknowledged influences of production and other acute concerns.

Disasters, man-made disasters theory said, are incubated. Prior to the disaster, there is a long period in which the potential for disaster builds. This period contains unnoticed or disregarded events that are at odds with the taken-for-granted beliefs about hazards and the norms for their avoidance. Turner considered managerial and administrative processes as the most promising for understanding this discrepancy between a build-up of risk and the sustained belief that it is under control. For Turner this space, gradually opening up between preserved beliefs and growing risk, was filled with human agency – perceptions, assessments, decisions, actions. Man-made disaster theory shifted the focus from engineering calculations of reliability to the softer, social side of failure. It shifted focus from structures onto processes, spread out over time, and over people and groups and organizations, all of them ironically tasked more or less with preventing the disaster from happening.

It was perhaps not surprising that the first book that took aim at the man-made origins of disaster came from Europe. And it is perhaps also not surprising that it remained virtually unknown in the U.S.A. for the next 15 years or more.[2] At the time, most social scientists interested in disasters were focusing on the pressing question of how disasters affect societies, and how societies could anticipate and parry those effects – absorb them, counteract them, ameliorate them, plan for them. Particularly in North America, disaster research in the 1950s, -60s and -70s was concerned mainly with the threat of nuclear apocalypse, and then mostly with managing its aftermath rather than exploring its origins. In fact, those origins were seen as geopolitical and technological in nature and therefore basically inaccessible to disaster research. As one result, there was little meaningful scholarly discussion of the possible sources of man-made disasters in other societal spheres.

THE INCUBATION AND SURPRISE OF FAILURE

The disaster incubation period was the most fascinating – to Turner, and to his successors. This is the phase, if you will, where "drift" happens. It is characterized by the "accumulation of an unnoticed set of events which are at odds with the accepted beliefs about hazards and the norms for their avoidance."[3] This discrepancy between the way the world is thought to operate and the way it really is, breeds slowly but surely, and … rarely develops instantaneously. Instead, there is an accumulation over a period of time of a number of events which are at odds with the picture of the world and its hazards represented by existing norms and beliefs. Within this "incubation period" a chain of discrepant event, or several chains of discrepant events, develop and accumulate unnoticed.[4]

Those beliefs collapse after a failure has become apparent (though, of course, a lot of psychological and political energy can go into attempts to preserve those beliefs).[5] Disasters, for Turner, were primarily sociological phenomena. They represent a disruption in how people believe their system operates; a breakdown of their own norms about risk and how to manage it. A year after publication of Turner's book, Stech applied this idea to the failure of Israeli intelligence organizations to foresee what would become known as the Yom Kippur war, even though all necessary data that pointed in that direction was available somewhere across the intelligence apparatus.[6] Zvi Lanir used the term "fundamental surprise," to capture this sudden revelation that one's perception of the world is entirely incompatible with what turned out to be the case.[7]

The developing vulnerability has long been concealed by the organization's belief that it has risk under control, a belief that it is entitled to according to its own model of risk and the imperfect organizational-cognitive processes that help keep it alive. This, according to Turner, is how a successful system produces failure as a normal, systematic by-product of its creation of success. The potential for failure does not build up at random, as if it were some abnormal, irrational growth alongside and independent from "normal" organizational processes. On the contrary, man-made disaster theory suggested how the potential for an accident accumulates precisely because the organization, and how it is configured in a wider administrative and political environment, is able to make opportunistic, non-random use of organized systems of production that until then are responsible for the organization's success. Dumping mining refuse on a tip just outside the coal mine, which necessarily abuts the village so that people can live close to their work, all collaborates in making the enterprise work and in keeping it economically sustainable without a lot of seemingly unnecessary transport of people and stuff. The whole arrangement seems unproblematic until a tip slide reveals that it isn't.

In this sense, Turner's account was a small but tantalizing preview of complexity – and systems thinking about disasters. It suggested how system vulnerability arises from the unintended and complex interactions between seemingly normal organizational, managerial and administrative features.[8] The sources of failure must be sought in the normal processes and relationships that make up organizational life, in other words, and not in the occasional malfunctioning of individual components.

Yet that is what the theory eventually did boil down to. Turner was fascinated by how a belief in a safe system (despite a gathering storm) is kept alive and gets reproduced. And his search for answers turned up broken parts: limits or problems in human cognition and communication. He found how people's erroneous assumptions helped explain why events went unnoticed or misunderstood. This, he said, had to do with rigidities of human belief and perception, and with the tendency to disregard complaints or warning signals from outsiders (who are not treated as credible or privy to inside knowledge about the system), as well as a reluctance on part of decision-makers at many levels to fear the worst outcome. He identified what he called "decoy phenomena," distractions away from the major hazard. For example, at Aberfan, residents mistakenly believed that the danger from tips was associated with the tipping of very fine waste, rather than more coarse surplus material. When it was agreed that fine waste would not be dumped on tips in the same way, residents withdrew some of their protests.

Turner also saw managerial and administrative difficulties in handling information in complex situations that blurred signal with noise. There were failures to comply with discredited or out-of-date regulations, and these passing unnoticed because of a cultural lag in what was accepted as normal. Then there was the "strangers and sites" problem: people (strangers) entering areas they officially shouldn't because of a lack of mandate or knowledge, and sites being used for purposes that was not their original intention. This all amounted to the judgment errors, cognitive lapses and communication difficulties that Turner saw as critical for creating the discrepancy in which failure was incubated. This point of view, focusing on which components went wrong where, is sustained in recent writings on high-reliability theory:

> Failure means that there was a lapse in detection. Someone somewhere didn't anticipate what and how things could go wrong. Something was not caught as soon as it could have been caught.[9]

This expresses the same Newtonian commitment as that which Turner could not escape. The gap between risk-in-the-world and risk-as-perceived grows because somebody, somewhere in the system, is not detecting things that could be detected. This represents broken components that can be tracked down and fixed or replaced.

RISK AS ENERGY TO BE CONTAINED: BARRIER ANALYSIS

Man-made disaster theory has left another theoretical legacy that has made it difficult to think of organizational risk in dynamic, adaptive terms and trace its development over time. Man-made disasters theory, from its very roots, takes for granted that we think of risk in terms of energy – a dangerous build-up of energy, unintended transfers, or uncontrolled releases of energy. This risk needs to be contained, and the most popular way is through a system of barriers: multiple layers whose function it is to stop or inhibit propagations of dangerous and unintended energy transfers. This separates the object-to-be-protected from the source of hazard by a series of defenses (which is a basic notion in the latent failure model). Other countermeasures include preventing or improving the recognition of the gradual build-up of dangerous energy (something that very much inspired man-made disaster theory), reduce the amount of energy (for example, reduce the height or composition of the tip), prevent the uncontrolled release of energy or safely distribute its release.

The conceptualization of risk as energy to be contained or managed has its roots in efforts to understand and control the physical nature of accidents. This also points to the limits of such a conceptualization. It is not necessarily well-suited to explain the organizational and socio-technical factors behind system breakdown, nor equipped with a language that can meaningfully handle processes of gradual adaptation, or the social processes of risk management and human decision-making. The central analogy used for understanding how systems work in this models is a technical system (for which the Newtonian–Cartesian worldview is optimally suited). And the chief strategy for understanding how these work and fail has always been reductionism. That means dismantling the system and looking at the parts that make up the whole. Consistent with Newtonian logic, this approach assumes that we can derive the macro properties of a system (for example, safety) as a straightforward combination or aggregation of the performance of the lower-order components or subsystems that constitute it. Indeed, the assumption is that safety can be increased by guaranteeing the reliability of the individual system components and the layers of defense against component failure so that accidents will not occur.

These assumptions are visible in one of the off-shoots of man-made disaster theory: the Swiss Cheese Model (also known as the latent failure model, or defenses in depth model), which was first published in the late 1980s.[10] It preserves the basic features of the risk-as-energy model. The Swiss Cheese Model relies on the sequential, or linear progression of failure(s) that became popular in the 1930s, particularly in industrial safety applications. There, adverse outcomes were viewed as the conclusion of a sequence of events. It was a simple, linear way of conceptualizing how events interact to produce a bad outcome. According to the sequence of events idea, events preceding the accident happen linearly, in a fixed order, and the accident itself is the last event in the sequence. In a slightly different version, it has been known, too, as the domino model, for its depiction of an accident as the endpoint in a string of falling dominoes.

The protection against failure is to put in barriers. These barriers, or defenses, need to be put in place to separate the object to be protected from the hazard. They are measures or mechanisms that protect against hazards or lessen the consequences of malfunctions or erroneous actions. These defenses come in a variety of forms. They can be engineered (hard) or human (soft), they can consist of interlocks, procedures, double-checks, actual physical barriers or even a line of tape on the floor of the ward (that separates an area with a particular anti-septic routine from other areas, for example). According to Reason, the "best chance of minimizing accidents is by identifying and correcting these delayed action failures (latent failures) before they combine with local triggers to breach or circumvent the system's defenses." This is consistent with ideas about barriers and the containment of energy or the prevention of uncontrolled release of energy.

But defense layers have "holes" in them. An interlock can be bypassed, a procedure can be ignored, a safety valve can begin to leak. An organizational layer of defense, for example, involves such processes as goal setting, organizing, communicating, managing, designing, building, operating, and maintaining. All of these processes are fallible, and produce the latent failures that reside in the system. This is not normally a problem, but when combined with other factors, they can contribute to an accident sequence. Indeed, according to the latent failure model, accidents happen when all of the layers are penetrated (when all their imperfections or "holes" line up). Incidents, in contrast, happen when the accident progression is stopped by a layer of defense somewhere along the way. The Swiss Cheese Model got its name from the image of multiple layers of defense with holes in them. Only a particular relationship between those holes, however (when they all "line up") will allow hazard to reach the object that was supposed to be protected.

It is interesting to see how people have, for a long time, modeled undesirable energy as proceeding along a linear trajectory that needs stopping or channeling. To this day, houses in Bali, Indonesia, are typically built according to specifications that keep evil spirits out. Evil spirits are believed to be able to travel in straight lines only, so if all the holes (for example, doors) line up, the spirit can enter the house, but will exit it again at the other end. Evil is thus released. If doors don't line up, evil stops traveling and is contained within the house.[11] The purpose of holes in the layers of defense (or doors in the walls) is opposite in these two ideas, of course. In the Swiss Cheese Model, holes that allow evil to travel through are bad. In the Balinese myth, they are good. But the notion of a linear trajectory along which bad influences on a system's health travel is the same. And, in the West, the notion of trajectories in three-dimensional space is of course entirely Newtonian.

As for the Swiss Cheese Model, the imperfection of the layers of defense can sponsor a search for broken parts (errors, communication difficulties, deficient supervision). Together with other theoretical assumptions, man-made disaster theory and its offshoots have thus retained a firmly Newtonian position about risk and people's knowledge of risk. Thinking of danger in terms of uncontrolled energy releases was one. The idea that there is a "real" risk out there, versus an "imagined" risk inside organizations and inside people's heads was another. This latter assumption made it difficult for accident theory to understand how people saw their world at the time and why this made sense to them. These assumptions probably helped keep the brakes on subsequent developments in accident research and would hamper progress on theorizing beyond the broken part. Today, a substantial gulf still lies between man-made disaster theory and its offshoots on the one end, and what is considered complexity- and systems thinking on the other.

Man-made disaster theory argues that (p. 16), "despite the best intentions of all involved, the objective of safely operating technological systems could be subverted by some very familiar and "normal" processes of organizational life."[12] Such "subversion" occurs through usual organizational phenomena such as information not being fully appreciated, information not correctly assembled, or information conflicting with prior understandings of risk. Turner noted that people were prone to discount, neglect or not take into discussion relevant information, even when available, if it mismatched prior information, rules or values of the organization.

The problem is that this doesn't really explain how or why people who manage a ward or a service or a hospital are unable to "fully" appreciate available information despite the good intentions of all involved. In the absence of such an explanation, the only prescription for them is to try a little harder, and to realize that safety should be their main concern. To try to imagine how risk builds up and travels through their organization. Indeed, man-made disaster theory offers that managerial and operational activities aimed at preventing a drift into failure both reflects and is promoted by at least the following four features:

○ Senior management commitment to safety
○ Shared care and concern for hazards and a willingness to learn and understand how they impact people
○ Realistic and flexible norms and rules about hazards
○ Continual reflection on practice through monitoring, analysis and feedback systems.

Through these four aspects, an organization is supposed to be able to continuously monitor weak signals and revise its responses to them. Some high-risk industries apparently succeeded in reaching an end state in which they were they were quite adept at this. Empirical observations of systems such as air traffic control, power generation and aircraft carriers in the 1980s grew into an entire school of thought – that of high-reliability organizations.

HIGH RELIABILITY ORGANIZATIONS

High reliability theory asks how risks are monitored, evaluated and reduced in organizations; what human actions, what deliberate processes lie behind that and how can they be enhanced? The theory argues that careful, mindful organizational practices can make up for the inevitable limitations on the rationality of individual members. High reliability theory describes the extent and nature of the effort that people at all levels in an organization can engage in to ensure consistently safe operations despite its inherent complexity and risks.[13]

During a series of empirical studies, high reliability organizational (HRO) researchers found that through leadership safety objectives, the maintenance of relatively closed systems, functional decentralization, the creation of a safety culture, redundancy of equipment and personnel, and systematic learning, organizations could achieve the consistency and stability required to achieve failure-free operations. Some of these findings seemed closely connected to the worlds studied – naval aircraft carriers, for example. There, in a relatively self-contained and isolated, closed system, systematic learning is an automatic by-product of the swift rotations of naval personnel, turning everybody into instructor and trainee, often at the same time. Functional decentralization meant that complex activities (like landing an aircraft and arresting it with the wire at the correct tension) were decomposed into simpler and relatively homogenous tasks, delegated down into small workgroups with substantial autonomy to intervene and stop the entire process independent of rank. High reliability researchers found many forms of redundancy – in technical systems, supplies, even decision-making and management hierarchies, the latter through shadow units and multi-skilling.

When researchers first set out to examine how safety is created and maintained in such complex systems, they focused on errors and other negative indicators, such as incidents, assuming that these were the basic units that people in these organizations used to map the physical and dynamic safety properties of their production technologies, ultimately to control risk. The assumption was wrong: they were not. Operational people, those who work at the sharp end of an organization, hardly defined safety in terms of risk management or error avoidance. Four ingredients kept reappearing, and they form the contours of what has become high reliability theory, or the theory of high reliability organizations.[14]

Leadership safety objectives are the first ingredient. Without such commitment, there is little point in trying to promote a culture of reliability. The idea is that others in the organization will never be enticed to find safety more important than their leadership. Short-term efficiency, or acute production goals, are openly (and sometime proudly) sacrificed when chronic safety concerns come into play. Agreement about the core mission of the organization (safety) is sought at every available opportunity, particularly through clear and consistent top-down communication about the importance of safety. Such commitments and communication may not be enough, of course. Other people have pointed out that the distance between loftily stated goals and real action is quite large, and can remain quite large, despite leadership or managerial pledges to the contrary.[15]

The need for redundancy is the second ingredient. The idea behind redundancy is that it is the only way to build a reliable system out of unreliable parts. Multiple and independent channels of communication and double-checks should, in theory, be able to produce a highly reliable organization. Redundancy in high reliability theory can take two forms: duplication and overlap. In duplication, two different units or people or parts perform the same function, often in real time. Duplication is also possible serially, as in the double-checking of a medication preparation. Overlap is redundancy where units or people or parts have some functional areas in common, but not all. It is obviously a cheaper solution.[16]

Decentralization, culture and continuity form the third ingredient. High reliability organizations rely on considerable delegation and decentralization of decision authority about safety issues. These organizations don't readily court government or regulatory interference with their activities and instead acknowledge the superiority of local entrepreneurial efforts to improve safety through engineering, procedure or training.[17] People inside organizations continually create safety through their evolving practice. In high-reliability organizations, active searching and exploration for ways to do things more safely is preferred over passively adapting to regulation or top-down control.

As a result, sharp-end practitioners in high-reliability organizations are entrusted to take appropriate actions in tight situations because they will have been inculcated through rituals, values, exercises and incentives. Having members work in a "total institution," isolated from wider society and inside their own world, seems to contribute to a culture of reliability. This aim is consistent with the maintenance of a relatively closed system. Finally, continuous operations and training, non-stop on-the-job education, a regular throughput of new students or other learners, and challenging operational workloads contribute greatly to reduced error rates and enhanced reliability.

Organizational learning is the fourth ingredient for high reliability organizations. High reliability grows out of incremental learning through trial and error. Things are attempted, new procedures or routines are tested (if carefully so), the effects are duly considered. Smaller dangers are courted in order to understand and forestall larger ones.[18] Simulation and imagination (for example, disaster exercises) are important ways of doing so when the costs of failure in the real system are too high. For high reliability theory, such learning does not need to be centrally orchestrated. In fact, the distributed, local nature of learning is what helps new and better ways of doing things emerge, a faster and quicker and more operationally grounded way of learning by trial and error.

CHALLENGING THE BELIEF IN CONTINUED SAFE OPERATIONS

Ensuing empirical HRO work, stretching across decades and a multitude of high-hazard, complex domains (aviation, nuclear power, utility grid management, navy) affirmed this picture. Operational safety – how it is created, maintained, discussed, mythologized – is much more than the control of negatives. As Gene Rochlin put it:

> The culture of safety that was observed is a dynamic, intersubjectively constructed belief in the possibility of continued operational safety, instantiated by experience with anticipation of events that could have led to serious errors, and complemented by the continuing expectation of future surprise.[19]

The creation of safety, in other words, involves a belief about the possibility to continue operating safely. This belief is built up and shared among those who do the work every day. It is moderated or even held up in part by the constant preparation for future surprise – preparation for situations that may challenge people's current assumptions about what makes their operation risky or safe. It is a belief punctuated by encounters with risk, but it can become sluggish by overconfidence in past results, blunted by organizational smothering of minority viewpoints, and squelched by acute performance demands or production concerns. Instead it should be a belief that is open to intervention so as to keep it curious, open-minded, complexly sensitized, inviting of doubt, and ambivalent toward the past.[20]

Note how these HRO commitments try to pull an organization's belief in its own infallibility away from Turner's disaster incubation. The past is no good basis for sustaining a belief in future safety, and listening to only a few channels of information will render the belief narrow and unchallenged. Disaster gets incubated when the organization's belief in continued safe operations is left to grow and solidify. So what can an organization do to continually challenge its own belief and to calibrate it against "reality"? How can any organization reach this desired high-reliability end state, and stay there?

The outlines of some answers appear in recent work on high reliability theory. It involves a preoccupation with failure, a reluctance to simplify, a sensitivity to operations, deference to expertise and a commitment to resilience.[21] Together, these commitments may be able to make drift visible before it becomes a full-blown failure or accident. Organizations that look relentlessly for symptoms of malfunctioning, and are able to link these symptoms to strategic positions or priorities, are better able to create practices that preclude problems or stop them from developing into unmanageably big ones. This still involves, of course, knowing what to look for (something that may have been really hard with the jackscrew in Chapter 2, as it is with any complex system). People in high-reliability organizations have been described as skeptical, wary and suspicious of quiet periods. And with good reason. Success, or the absence of symptoms of danger:

> Breeds confidence and fantasy. When an organization succeeds, its managers usually attribute success to themselves or at least to their organization, rather than to luck. The organization's members grow more confident of their own abilities, of their manager's skills, and of their organization's existing programs and procedures. They trust the procedures to keep them apprised of developing problems, in the belief that these procedures focus on the most important events and ignore the least significant ones.[22]

> Success narrows perceptions, changes attitudes, reinforces a single way of doing business, breeds overconfidence in the adequacy of current practices, and reduces the acceptance of opposing points of view.[23]

Pre-occupation with failure is not just a commitment to detecting anything that might signal a malfunction. It also is an active consideration of all the places and moments where you don't want to fail. Ongoing attention to traps and pitfalls is ingrained in how high-reliability organizations work. The reluctance to simplify is part of this, too. Rather than lumping events into large categories (human error or a maintenance squawk, not a flight safety problem), high reliability organizations try to pay close attention to the context and contingencies of such events. More differentiation, more worldviews, more mindsets come into view that way. High-reliability theory suggests that it is this complexity of possible interpretations of events that allows organizations to better anticipate and detect what might go wrong in the future. Accidents such as those with the *Challenger* and *Columbia* Space Shuttles have shown that a lack of deference to expertise and sensitivity to actual system operations (that is, prioritizing managerial views rather than operational ones) contributes to organizational brittleness. People who were not most technically qualified to make decisions about the continued operation of the system got to make those decisions anyway. Expert assistance was not easily available during important meetings. And where it was, it was made subservient to larger system goals (for example, production, launch schedules). These aspects help obscure any signs of drifting into failure, and high-reliability theory suggests organizations should take expertise seriously, listen to minority viewpoints and remain less concerned with strategy and more sensitive to daily operations.[24]

To be sure, even a deference to expertise and a sensitivity to operations does not insulate a system from the possibility of drifting into failure. After all, expert opinion too is based on uncertain, incomplete knowledge, because most technology is unruly. Often it is not clear even to most to insiders that drift toward margins is occurring as a result of their decisions or their managers' priorities, or it isn't obvious that it matters if it is. Even when it is clear that drift toward margins is occurring, the consequences may be hard to foresee, and judged to be a small potential loss in relation to the immediate gains. Experts do their best to meet local conditions, and in the busy daily flow and complexity of activities they may be unaware of any potentially dangerous side effects of their or their managers' decisions. It is only with the benefit of hindsight or omniscient oversight (which is utopian) that these side effects can be linked to actual risk. Some, then, see the high-reliability call to defer to expertise as limited:

> We should not expect the experts to intervene, nor should we believe that they always know what they are doing. Often they have no idea, having been blinded to the situation in which they are involved. These days, it is not unusual for engineers and scientists working within systems to be so specialized that they have long given up trying to understand the system as a whole, with all its technical, political, financial, and social aspects. (p. 368)[25]

Being an insider of a system, then, can make systems thinking all but impossible. It can make it really hard to transcend the daily noise and seeming importance of decisions and deadlines, with all their "political, financial, and social aspects." Influences from outside the technical knowledge base can exert a subtle but powerful pressure on the decisions and trade-offs that people make, and constrain what is seen as a rational decision or course of action at the time. Thus, even though experts may be well-educated and motivated, even to them a "warning of an incomprehensible and unimaginable event cannot be seen, because it cannot be believed"[26] This places severe limits on the rationality that can be brought to bear in any decision-making situation: "seeing what one believes and not seeing that for which one has no beliefs are central to sensemaking. Warnings of the unbelievable go unheeded."[27] That which cannot be believed will not be seen. Even if relevant events and warnings end up in a reporting system (which is doubtful because they are not seen as warnings even by those who would do the reporting), it is even more generous to assume that further expert analysis of such incident databases could succeed in coaxing the warnings into view.

The difference, then, between insight at the time and hindsight (after an accident) is tremendous. With hindsight, the internal workings of the system may become lucid: the interactions and side effects are rendered visible. And with hindsight, people know what to look for, where to dig around for the rot, the missing connections. Triggered by the Alaska 261 accident, the regulator launched a special probe into the maintenance-control system at Alaska Airlines. It found that procedures in place at the company were not followed, that controls in place were clearly not effective, that authority and responsibility were not well defined, that control of the maintenance-deferral systems was missing, and that quality-control and quality-assurance programs and departments were ineffective. It also found incomplete C-check paperwork, discrepancies of shelf-life expiration dates of parts, a lack of engineering approval of maintenance work-card modifications and inadequate tool calibrations. Maintenance manuals did not specify procedures or objectives for on-the-job training of mechanics, and key management positions (for example, safety) were not filled or did not exist. Indeed, constraints imposed on other organizational levels were nonexistent, dysfunctional, or eroded.

But as has been pointed out previously, seeing holes and deficiencies in hindsight is not an explanation of the generation or continued existence and rationalization of those deficiencies. It does not help predict or prevent a drift into failure. Instead, the processes by which such decisions come about, and by which decision-makers create their local rationality, are one key to understanding how safety can erode on the inside a complex, sociotechnical system. Why did these things make sense to organizational decision-makers at the time? Why was it all normal, why was it not reportworthy, not even for the regulator tasked with overseeing these processes?

Picking up the example of Alaska 261 from Chapter 1, new maintenance guidance for the MD-80 series aircraft (the so-called MSG-3 MD-80 MRB) decided that the 3,600 hour jackscrew lubrication task change should be a part of the larger C-check package.[28] The review board that decided this did not consult the manufacturer's design engineers, nor did it make them aware of the extension of the maintenance interval that would result from their decision. The manufacturer's initial maintenance document for the DC-9 and MD-80 lubrication, specifying an already extended 600- to 900-hour interval (departing from the 1964 recommendation for 300 hours), was also not considered by the board developing the new maintenance guidance.

From a local perspective, with the pressure of time limits and constraints on available knowledge, the decision to extend the interval without adequate expert input must have made sense. People consulted at the time must have been deemed adequate and sufficiently expert in order to feel comfortable enough to continue. The creation of rationality must have been seen as satisfactory. Otherwise, it is hard to believe that board would have proceeded as it did. But the eventual side effects of these smaller decisions were not foreseen. From the larger perspective, the gap between production and operation, between making and maintaining a product, was once again allowed to widen. A relationship that had been instrumental in helping bridge that gap (consulting with the original design engineers who make the aircraft, to inform those who maintain it), a relationship from history to (then) present, was severed.

If not foreseeing side effects made sense for the review board (and this may well have been a banal result of the sheer complexity and paper load of the work mandated), it may make sense for participants in any other socio-technical system too. These decisions are sound when set against local judgment criteria; given the time and budget pressures and short-term incentives that shape behavior. Given the knowledge, goals, and focus of attention by the decision-makers, as well as the nature of the data available to them at the time, it made sense. It is in these normal, day-to-day processes, where we can find the seeds for drifting into failure. A very important ingredient in these processes is the existence of production goals, resource scarcity or throughput pressure. This creates goal interactions that ask people to sacrifice one thing against another. They may not do so consciously. In fact, the pattern of drift into failure suggests that the influences of those factors gradually becomes pre-rational, a taken-for-granted part of the worldview that decision-makers bring to work with them. Let's turn to how this might work with goal conflicts and the internalization of production pressure.

GOAL INTERACTIONS AND PRODUCTION PRESSURE

Turner, the man behind man-made disaster theory, implicitly raised a question that had been at the forefront of sociological thought for quite a while: what is "normal"? Is it "normal" to have a tip right outside a village? Is it acceptable? It apparently was, otherwise people would not have continued with it. The danger could not be seen, for people didn't believe it existed. Or the people who had the power to do something about it didn't believe it existed (or could not afford to admit it), because it would have interfered tremendously with getting the mining job done productively and efficiently. One of the ingredients in almost all stories of drift is a focus on production and efficiency. Pressures to achieve those goals can be felt acutely, and the effect of managerial or operational decisions on the ability to achieve them can often be measured directly. How these same decisions obscure chronic safety concerns is not easy to quantify at all. How is it that global pressures of production and scarcity find their way into local decision niches, and how is it that they there exercise their often invisible but powerful influence on what people think and prefer; what people then and there see as rational or unremarkable?

The answer may lie in what sociologists call the *macro-micro connection*: the link between macro-level forces that operate on an entire organization or system, and the micro-level cognitions and decisions of individual people in it. This link is far from straightforward. One starting point for understanding it is to trace the organization's goals and how they might conflict. An organization that tries to achieve multiple goals at the same time means goal conflicts for those working inside it. People will routinely encounter basic incompatibilities in what they need to strive for in their work. That in itself is nothing remarkable or unusual. As Dietrich Dörner observed, "Contradictory goals are the rule, not the exception, in complex situations."[29] This is the essence of most operational systems. Though safety is a (stated) priority, operational systems do not exist to be safe. They exist to provide a service or product, to achieve economic gain, to maximize capacity utilization. But still they have to be safe (in some sense, safety, or at least an image of safety, is a precondition for achieving any of the other goals). If we want to understanding drift into failure, we have to be particularly interested in how people themselves view these conflicts from inside their operational reality, and how this contrasts with management (and regulator) views of the same activities.

NASA's "Faster, Better, Cheaper" organizational philosophy in the late nineties epitomized how multiple, contradictory goals are simultaneously present and active in complex systems.[30] The loss of the Mars Climate *Orbiter* and the Mars *Polar Lander* in 1999 were ascribed in large part to the irreconcilability of the three goals (faster and better and cheaper) which drove down the cost of launches, made for shorter, aggressive mission schedules, eroded personnel skills and peer interaction, limited time, reduced the workforce, and lowered the level of checks and balances normally found (NASA 2000). People argued that NASA should pick any two from the three goals. Faster and cheaper would not mean better. Better and cheaper would mean slower. Faster and better would be more expensive. Such reduction, however, obscures the actual reality facing operational personnel in safety-critical settings. These people are there to pursue all three goals simultaneously – fine-tuning their operation, as Starbuck and Milliken say, to "render it less redundant, more efficient, more profitable, cheaper, or more versatile."[31] Fine-tuning , in other words, to make it faster, better, cheaper.

The 2003 Space Shuttle *Columbia* accident focused attention on the maintenance work that was done on the Shuttle's external fuel tank, once again revealing the differential pressures of having to be safe and getting the job done (better, but also faster and cheaper). A mechanic working for the contractor, whose task it was to apply the insulating foam to the external fuel tank, testified that it took just a couple of weeks to learn how to get the job done, thereby pleasing upper management and meeting production schedules. An older worker soon showed him how he could mix the base chemicals of the foam in a cup and brush it over scratches and gouges in the insulation, without reporting the repair. The mechanic soon found himself doing this hundreds of times, each time without filling out the required paperwork. Scratches and gouges that were brushed over with the mixture from the cup basically did not exist as far as the organization was concerned. And those that did not exist could not hold up the production schedule for the external fuel tanks. Inspectors often did not check. A company program that once had paid workers hundreds of dollars for finding defects had been watered down, virtually inverted by incentives for getting the job done now.

In the cases above, what was normal versus what was deviant was no longer so clear. Were these signals that pointed to a drift into failure? Goal conflicts between safer, better and cheaper were reconciled by doing the work more cheaply, superficially better (brushing over gouges in the foam covering the external tank of the Space Shuttle) and apparently without cost to safety. As long as Orbiters kept coming back safely, the contractor must have been doing something right. Understanding the potential side effects was very difficult given the historical mission success rate. Lack of failures were seen as a validation that current strategies to prevent hazards were sufficient. Could anyone foresee, in a vastly complex system, how local actions as trivial as brushing chemicals from a cup could one day align with other factors to push the system over the edge? Recall Weick's and Perrow's warning: what cannot be believed cannot be seen. Past success could comfortably be taken as a guarantee of continued safety.

Of course, some organizations pass some of their goal conflicts on to individual practitioners quite openly. There are airlines, for example, that pay their crews a bonus for on-time performance. An aviation publication commented on one of those operators (a new airline called Excel, flying from England to holiday destinations):

> As part of its punctuality drive, Excel has introduced a bonus scheme to give employees a bonus should they reach the agreed target for the year. The aim of this is to focus everyone's attention on keeping the aircraft on schedule.[32]

Most important goal conflicts, however, are never made so explicit. Rather, they are left to emerge from multiple irreconcilable directives from different levels and sources, from subtle and tacit pressures, from management or customer reactions to particular trade-offs. Organizations often resort to "conceptual integration, or plainly put, doublespeak."[33]

For example, the operating manual of one airline opens by stating that "(1) our flights shall be safe; (2) our flights shall be punctual; (3) our customers will find value for money." Conceptually, this is Dörner's doublespeak; documentary integration of incompatibles. It is impossible, in principle, to do all three simultaneously, as with NASA's faster, better, cheaper.

While incompatible goals arise at the level of an organization and its interaction with its environment, the actual managing of goal conflicts under uncertainty gets pushed, or trickles down into local operating units – control rooms, patient wards, airline cockpits, and the like. There the conflicts are to be negotiated and resolved in the form of thousands of little and larger daily decisions and trade-offs. These are no longer decisions and trade-offs made by the organization, but by individual operators or crews. It is this insidious delegation, this handover, where the internalization of external pressure takes place. Global tension between efficiency and safety seeps into local decisions and trade-offs by individual people or groups – the macro-micro connection. When do the problems and interests of an organization under pressure of resource scarcity and competition, become the problems and interests of individual actors at several levels within that organization? Experts and practitioners inside safety-critical organizations routinely make the reconciliation of goal conflicts their own problem. They typically want to be on time. They want to deliver a service or product that is value for money, they want to be more efficient, achieve greater levels of quality. So perhaps practitioners do pursue incompatible goals of faster, better and cheaper all at the same time and are aware of it too. In fact, practitioners take their ability to reconcile the irreconcilable as source of considerable professional pride. In many worlds, it is seen as a strong sign of their expertise and competence.

Crews of one airline describe their ability to negotiate these multiple goals while under the pressure of limited resources as "the blue feeling" (referring to the dominant color of their fleet).[34] This "feeling" represents the willingness and ability to put in the work to actually deliver on all three goals simultaneously (safety, punctuality and value for money). This would confirm that practitioners do pursue incompatible goals of faster, better and cheaper all at the same time and are aware of it, too. The "blue feeling" signals crews' strong identification with their organization and its brand. Yet it is a feeling only individuals or crews can have; a feeling because it is internalized. Insiders point out how some crews or commanders have the blue feeling while others do not. It is a personal attribute, not an organizational property. Those who do not have the blue feeling are marked by their peers – seldom supervisors – for their insensitivity to, or disinterest in, the multiplicity of goals; for their unwillingness to do substantive cognitive work necessary to reconcile the irreconcilable. These practitioners do not reflect the corps' professional pride since they will always make the easiest goal win over the others (for example, "don't worry about customer service or capacity utilization, it's not my job"), choosing the path of least resistance and least work in the eyes of their peers. In the same airline, those who try to adhere to minute rules and regulations are called "Operating Manual worshippers" – a clear signal that their way of dealing with goal contradictions is not only perceived as cognitively cheap (just go back to the book, it will tell you what to do), but as hampering the collective ability to actually get the job done, diluting the blue feeling.

Thus macro-structural forces, that operate on an entire organization, get expressed in how local work groups generate assessments about risk and make implicit statements about what they see as professionalism. Macro-level pressures are reproduced or manifested in what individual people or groups do. Their achievements in the face of goal conflicts doesn't just represent a personal attribute. It also becomes an inter-peer commodity that affords comparisons, categorizations and competition among members of the peer group, independent of other layers or levels in the organization. This can be seen to operate as a subtle engine behind the negotiation among different goals in safety-critical professions such as airline pilots, flight dispatchers, air traffic controllers, emergency department managers or aircraft maintenance workers.[35] Aircraft maintenance has incorporated even more internal mechanisms to deal with goal interactions. The demand to meet technical requirements clashes routinely with time- or other resource-constraints such as inadequate time, personnel, tools, parts, or functional work environment (McDonald et al. 2002). The vast internal, sub-surface networks of routines, illegal documentation and "shortcuts," which from the outside would be seen as massive infringement of existing procedures, are a result of the pressure to reconcile and compromise.

The same could be seen in the maintenance and repair of Space Shuttle fuel tanks by the contractor Lockheed. There, too, it was the actual work practices (and the efficiency achieved through them) that constituted the basis for technicians' strong professional pride and sense of responsibility. They were able to deliver safe work, apparently even exceeding technical requirements (all evidence of damage was gone, after all). Seen from the inside, it is the role of the technician to apply judgment founded on his or her knowledge, experience and skill – not on formal procedure. Those most adept at this are highly valued for their productive capacity even by higher organizational levels.[36]

NORMALIZING DEVIANCE, STRUCTURAL SECRECY AND PRACTICAL DRIFT

Two landmark studies, one on the launch of the Space Shuttle *Challenger*,[37] the other on the shooting down of two Black Hawk helicopters over northern Iraq in 1993,[38] have delved deeply into the sociology of decision-making and the resulting organizational drift. The first one has taken a completely new look at the role of production pressure and how it influences organizational rationality in the face of goal conflicts between safety and schedules. Vaughan's study, more than any other, explored in great depth the macro-micro connection and described how production pressures became part and parcel not just of managerial decision-making, but of engineering judgment and risk assessment. She was able to shed particular light on processes of organizational adaptation and the flexibility of social interpretations of risk in the face of uncertainty. With her work, she began to fill in the blanks of Barry Turner's incubation period, replacing Turner's broken parts (communication errors, managerial misjudgments, inflexible beliefs) with a much more nuanced and socially patterned story. What people did inside this hugely complex and disjointed organization made sense to them at the time. Each little decision they made, each little step, was only a small step away from what had previously be seen as acceptable risk. And each safe Space Shuttle return proved that they had been right.

THE NORMALIZATION OF DEVIANCE

In 1986, the Space Shuttle *Challenger* broke apart 73 seconds after the launch of its tenth mission, resulting in the death of all seven crew members, including a civilian teacher. The accident was investigated by a Presidential Commission and made to fit a structural account. Organizational deficiencies (holes in the layers of defense) had led to a tendency for engineers and managers not to report upward. These holes needed to be fixed by changes in personnel, organization, or indoctrination. The report placed responsibility for communication failures with middle managers, who had overseen key decisions and the implementation of inadequate rules and procedures. The Presidential Commission traced their managerial wrongdoing to a combination of enormous economic strains *and* operational expectations put on the Space Shuttle.

Space Shuttles had been designed and built (and sold to the tax payer) as a bus-like recyclable and operational technology for trips into space. In reality, production expectations and operational capacities were very far apart. In order to try to live up to launch schedules despite innumerable technical difficulties, middle management allowed rule violations and contributed to the silencing of those with bad news. Information about hazards (among others, an O-ring blow-by problem in solid rocket boosters that eventually led to the *Challenger* accident) was suppressed in order to stick to the launch schedule. On the eve of the launch of *Challenger*, managers consciously decided to take the risk of launching a troubled design in exceptionally low temperature. A teacher was going to be on board, and the president was going to hold his State of the Union address during her trip in space. The Presidential Commission exonerated top administrators, arguing that they did not have the information. It blamed middle management, however, because they did have all the information. They were warned against the launch by their engineers, but decided to proceed anyway.

The Presidential Commission's account was consistent with rational choice theory, in which humans are considered to be perfectly rational decision-makers. NASA middle management was assumed to have had exhaustive access to information for their decisions, as well as clearly defined preferences and goals about what they want to achieve. Despite this, they decided to go ahead with the launch anyway, to gamble. They were amoral calculators, in other words.

Richard Feynman, physicist and maverick member of the Presidential Commission, wrote a separate, dissenting appendix to the report on his own conclusion. One issue that disturbed him particularly was the extent of structural disconnects across NASA and contractor organizations. These uncoupled engineers from operational decision-making and led to an inability to convince managers of (even the utility of) an understanding of fundamental physical and safety concepts. Professional accountability and the dominance of a technical culture and expertise had gradually been replaced with bureaucratic accountability, where administrative control is centralized at the top, and where the focus of decision-makers is trained on business ideology and the meeting of political expectations.

But middle managers would not have possessed everything to base rational decisions on. People do not have full knowledge of all relevant information, possible outcomes, relevant goals. It would be impossible. What they do instead makes sense to them given their local goals, their current knowledge and their focus of attention at the time. So, Diane Vaughan argued, if there was amoral calculation and conscious choice to gamble on the part of middle management, then it was so only in hindsight.

It wasn't difficult to agree on the basic facts. NASA had been under extraordinary production pressure and economic constraints in the years leading up to the accident. Its reusable spacecraft had been sold on the promise of cost effectiveness, and won an endorsement from the Air Force (which put outlandish payload expectations on the design in return). In order to break even, it had to make 30 flights per year, and even this was considered a conservative estimate. Funds eventually allocated to NASA were half of what had been requested. NASA never got all the funds it needed, and the Space Shuttle was moved from a developmental and testing flight regime into an operational one early on. The design was one big trade-off. Reductions in development costs that were needed to secure funding were attained by exporting cost into higher future operating costs. True launch costs ballooned up to 20 times the original estimates, even as competition was growing from commercial spaceflight in other countries and a third of the NASA workforce had disappeared as compared to the *Apollo* years. Only nine missions were flown in 1985, the year before *Challenger*, and even the launch of *Challenger* was delayed multiple times.

It was easy to reach agreement on the notion that production pressure and economic constraints are a powerful combination that could lead to adverse events. But how? Was it a matter of conscious trade-offs, of actively sacrificing safety and making immoral calculations that gambled with people's lives? Or was something else going on? Researchers started to dig into the social processes that slowly but surely converted the bad news of technical difficulties and operational risk into something that was normal, acceptable, expected, manageable.

These are important questions at the heart of drift into failure. If managers know that technologies have problems, are they irrational and amoral in implementing them anyway? And are expert practitioners powerless and disenfranchised in the face of their decisions, like NASA's engineers in the 1980s? To the Presidential Commission there was a linear relationship between scarcity, competition, production pressure and managerial wrongdoing. Pressures developed because of the need to meet customer commitments, which translated into a requirement to launch a certain number of flights per year and to launch them on time. Such considerations may occasionally have obscured engineering problems. Vaughan traced the subtle evolution of risk assessments by those involved with the key technology in *Challenger* (the solid rocket boosters). She showed how insiders continuously negotiated the meaning of damage to the solid rocket booster O-rings as problems with the design unfolded.

For Vaughan, seeing holes and deficiencies in hindsight was not an explanation of the generation or continued existence of those deficiencies. It wouldn't help predict or prevent failure. Instead, the processes by which such decisions come about, and by which decision-makers create their local rationality, are one key to understanding how safety can erode on the inside a complex, socio-technical system. Why did these things make sense to organizational decision-makers at the time? Why was it all normal, why was it not worthy of reporting?

Of course, production pressures played a huge role, but not in the way envisioned by the Presidential Commission. Rather, production pressures and resource limitations gradually became institutionalized, taken for granted. They became a normal aspect of the worldview every participant brought to organizational and individual operational decisions. To Vaughan, this social-technical dynamic, where the idea of "risk" is continuously constructed and renegotiated at the intersection of the technical and the social, was at the heart of the incubation period for *Challenger*. Indeed, this dynamic is fundamental to the construction of risk in any safety-critical endeavor (Vaughan 1999). She called it the normalization of deviance. This is how a workgroup's construction of risk can persist even in the face of continued and worsening signals of potential danger. Signals of potential danger on one flight are acknowledged and then rationalized and normalized, leading to a continued use under apparently similar circumstances. This repeats itself until something goes wrong, revealing the gap between how risk was believed to be under control and its actual presence in the operation:

○   *Beginning the construction of risk: a redundant system.* The starting point for the safety-critical activity is the belief that safety is assured and risk is under control. Redundancies, the presence of extraordinary competence, or the use of proven technology can all add to the impression that nothing will go wrong. There is a senior surgeon who watches the operation through a laparoscope from a distance.

○   *Signals of potential danger.* Actual use or operation shows a deviation from what is expected. This creates some uncertainty, and can indicate a threat to safety, thus challenging the original construction of risk.

○   *Official act acknowledging escalated risk.* Evidence is shown to relevant people, a meeting may be called.

○   *Review of the evidence.* After the operation, discussions may ensue about who did what and how well things went. This is not necessarily the case after all operations; in fact, it could be the kind of standardized debrief that is hardly institutionalized in medicine yet (see Chapter 6).

○   *Official act indicating the normalization of deviance: accepting risk.* The escalated risk can get rationalized or normalized as in-family, as expected. It may be argued that redundancy was assured at multiple levels in the system. And, after all, the technology itself has undergone various stages of testing and revision before being fielded. All these factors contribute to a conclusion that any risk is duly assessed, and under control.

○   *Continued operation.* The technology will be used again in the next flight, because nothing went wrong, and a review of the risks has revealed that everything is under control.

The key to the normalization of deviance is that this process, this algorithm, repeats itself. And that successful outcomes keep giving the impression that risk is under control. Success typically leads to subsequent decisions, setting in motion a steady progression of incremental steps toward greater risk. Each step away from the previous norm that meets with empirical success (and no obvious sacrifice of safety) is used as the next basis from which to depart just that little bit more again. It is this decrementalism that makes distinguishing the abnormal from the normal so difficult. If the difference between what "should be done" (or what was done successfully yesterday) and what is done successfully today is minute, then this slight departure from an earlier established norm is not worth remarking or reporting on. decrementalism is about continued normalization: It allows normalization and rationalizes it.

Experience generates information that enables people to fine-tune their work: fine-tuning compensates for discovered problems and dangers, removes redundancy, eliminates unnecessary expense, and expands capacities. Experience often enables people to operate a socio-technical system for much lower cost or to obtain much greater output than the initial design assumed.[39]

Normalizing deviance is about fine-tuning, adaptation, and increments. Decisions that are seen as "bad decisions" after an adverse event (like using unfamiliar or non-standard equipment) seemed like perfectly good or reasonable proposals at the time. No amoral calculating people were necessary to explain why things eventually went wrong. All it took was normal people in a normal organization under the normal sorts of pressures of resource constraints and production expectations. People with normal jobs and real constraints but no ill intentions, whose constructions of meaning co-evolved relative to a set of environmental conditions, and who tried to maintain their understanding of those conditions. Because of this, says Vaughan,

> Mistake, mishap, and disaster are socially organized and systematically produced by social structures. No extraordinary actions by individuals explain what happened: no intentional managerial wrongdoing, no rule violations, no conspiracy. The cause of the disaster was a mistake embedded in the banality of organizational life and facilitated by an environment of scarcity and competition, an unprecedented, uncertain technology, incrementalism, patterns of information, routinization, organizational and interorganizational structures ...[40]

Vaughan's analysis shows how production pressures and resource constraints, originating in the environment, become institutionalized in the practice of people. These pressures and constraints have a nuanced and unacknowledged yet pervasive effect on organizations and the decision-making that goes on in them. This is why harmful outcomes can occur in activities and organizations constructed precisely to prevent them (indeed, like healthcare). The structure of production pressure and resource scarcity becomes transformed into organizational mandates, and affects what individual people see as normal, as rational, as making sense at the time.

There is a relentless inevitability of mistake, the result of the emergence of a can-do culture. This is created as people interact in normal work settings where they normalize signals of potential danger so that their actions become aligned with organizational goals. This normalization ensures that they can tell themselves and others that no undue risk was taken, and that in fact the organization benefited: the operation was completed, no resources were used unnecessarily or wasted. Through repeated success, this work group culture persists; it rationalizes past decisions while shaping future ones. But in the end, the decrementalism of those decisions contributes to extraordinary events.

HOW TO PREVENT THE NORMALIZATION OF DEVIANCE

Vaughan is not optimistic. Adverse events are "produced by complicated combinations of factors that may not congeal in exactly the same way again."[41]

Conventional explanations that focus on managerial wrongdoing leave a clear recipe for intervention (indeed the one supplied by the Presidential Commission). Get rid of the really bad managers, change personnel around, amend procedures and rules. But it is unclear whether this might prevent a repetition of processes of normalization:

> We should be extremely sensitive to the limitations of known remedies. While good management and organizational design may reduce accidents in certain systems, they can never prevent them ... system failures may be more difficult to avoid than even the most pessimistic among us would have believed. The effect of unacknowledged and invisible social forces on information, interpretation, knowledge, and – ultimately – action, are very difficult to identify and to control.[42]

Indeed, the Space Shuttle *Columbia* accident in 2003 showed how NASA history was able to repeat itself: foam strikes to the wing of the Space Shuttle had become normalized as acceptable flight risk, and converted into a maintenance problem rather than a flight-safety problem.

Vaughan's analysis reveals a more complex picture behind the creation of such accidents, and the difficulty in learning from them. It shifts attention from individual causal explanations to factors that are difficult to identify and untangle, yet have a great impact on decision-making in organizations. Vaughan's is a more frightening story than the historically accepted interpretation. Its invisible processes and unacknowledged influences tend to remain undiagnosed and therefore elude remedy.

The solution to risk, if any, is to ensure that the organization continually reflects critically on and challenges its own definition of "normal" operations, and finds ways to prioritize chronic safety concerns over acute production pressures. But how is this possible when the definition of "bad news" is something that gets constantly renegotiated, as success with improvised procedures or imperfect technology accumulates? Such past success is taken as guarantee of future safety. Each operational success achieved at incremental distances from the formal, original rules or procedures or design requirements can establish a new norm. From there a subsequent departure is once again only a small incremental step.

STRUCTURAL SECRECY AND PRACTICAL DRIFT

Large organizations like those necessary to maintain aircraft or launch spacecraft are made up of an enormous number of levels, departments, disciplines, and specializations. In the wake of an adverse event (particularly one that generates media attention and becomes a celebrated case), it is easy to assert that some people may have intentionally concealed information about the difficulties with a particular person or technology, thus preventing the administration from intervening and addressing the problem. But that is often only obvious in hindsight. Such secrecy is often simply a structural by-product of how work is organized, particularly inside large, bureaucratic organizations.

March, Cohen and Olsen developed an important perspective on the functioning of people inside such organizations.[43] They saw organizations as natural, open systems, not rational, closed systems in which people can accomplish work in a way that is immune from everything that makes us human. They are "natural" like all social groups. People actively pursue goals of narrow self-interest, and may prioritize things that only benefit their local group. This can well be to the detriment of official goals of openness or profit or production. The idea that organizations are capable of inculcating a safety orientation among its members through recruitment, socializing and indoctrination is met with great skepticism. Then, the organizations in which this happens are not closed off to the environment. They are "open" in the sense that they are constantly perfused by forces from outside: political, economic, social. These forces enter into all kinds of decisions and preferences that get expressed by various groups and people on the inside.

March and colleagues argued that organizations cannot make up for the limited, local rationalities of the people inside it. Instead, organizations themselves have a very limited, or local rationality because they are made up of people put together. Problems are ill-defined and often unrecognized, decision-makers have limited information and shifting allegiances and uncertain intentions. Solutions may be lying around, actively searching for a problem to attach themselves to. Possible moments for rational choice are opportunistic and badly coordinated. People often hardly prepare well for these moments (for example, budget or board meetings) yet the organization is expected to produce behavior that can be called a decision anyway.

This vision of organizational life stands in contrast with high reliability theory, which, as you might recall, argues that organizations in fact can make up for the local rationalities of the people inside it. It just takes better organization. But for March and colleagues organizations operate on the basis of a variety of inconsistent and ill-defined preferences. This leads to goal conflicts (even shifting and changing goal conflicts) and diversions of aims and direction across various departments. The organization may not even know its preference until after a decision has been made. Also, organizations use of unruly technology in their operations means that people inside might not even really understand while some things work and some things don't. Finally, organizational life is characterized by an extremely fluid participation in decision-making processes. People come and go, with various extents of commitment to organizational vision and goals. Some pay attention while they are there, others do not. Key meetings may be dominated by biased, uninformed or even uninterested personnel.

With this vision, it is not surprising that an organization is hardly capable of arriving at a clear understanding of its risks, and commitments for how they should be managed. Structural secrecy, with participants not knowing about what goes on in other parts of the organization, is a normal by-product of the bureaucratic organization and social nature of complex work. By structural secrecy, Vaughan meant the way that patterns of information, organizational structure, processes and transactions all undermined people's attempts to understand what was going on inside the organization. As organizations grow larger and larger, most actions are no longer directly observable. Labor is divided between subunits, hierarchies, and geographically dispersed. Bad news can remain localized, seen as normal where it is, and not get subjected to any credible outside scrutiny. Distance, both physical and social, interferes with efforts from those at the top to know what is going on and where (Vaughan 1996):

> Specialized knowledge further inhibits knowing. People in one department or division lack the expertise to understand the work in another or, for that matter, the work of other specialists in their own unit ... Changing technology also interferes with knowing, for assessing information requires keeping pace with these changes – a difficult prospect when it takes time away from one's primary job responsibilities. To circumvent these obstacles, organizations take steps to increase the flow of information – and hypothetically, knowledge. They make rules designating when, what, and to whom information is to be conveyed. Information exchange grows more formal, complex, and impersonal, perhaps overwhelmingly so, as organizations institute computer transaction systems to record, monitor, process and transmit information from one part of the organization to another. Ironically, efforts to communicate more can result in knowing less.[44]

People do what locally makes sense to them, given their goals, knowledge and focus of attention in that setting. Scott Snook, who produced a revisionist account of the mistaken shooting down of two U.S. Black Hawk helicopters by U.S. fighter jets over northern Iraq in April 1994, called it "practical action." This is "behavior that is locally efficient, acquired through practice, anchored in the logic of the task, and legitimized through unremarkable repetition."[45] The incident studied by Snook started with Operation Provide Comfort (OPC), which followed the establishment a no-fly zone for Iraqi warplanes over northern Iraq because of a feared response by Saddam Hussein to a Kurdish uprising there in 1991. A United Nations resolution passed in 1991 called on Iraq to end its repression of the Kurds in its north and opened the way for the delivery of relief aid to the Kurds. North of the 36th parallel, U.S., U.K. and French aircraft patrolled the no-fly zone, so as to keep the airspace free of Iraqi intruders and open for relief efforts.

Operation Provide Comfort had been driven by a specialized Task Force. Over time, however, the many local units that worked under the OPC umbrella had developed their own logics and procedures and routines, away (and differentially so) from the centrally dictated ones. This is normal, as operational people discover that the original rules don't fit most situations most of the time. So they act in ways that align better with their perception of the demands at the time. Rules become less important in shaping action, task demands and environmental changes become more important. Snook called the phenomenon "practical drift."

The day of the shooting down fatefully brought together a number of different units, from helicopters to fighter jets to AWACS aircraft overhead to equipment ill-suited for the task, as well as different command structures (European and U.S.). The consequences of practical drift became evident as the system momentarily contracted into a very tightly coupled situation, in which recovery was very difficult.

As with Vaughan's normalization of deviance, operational success with such adapted procedures is one of the strongest motivators for doing it again, and again. Recall from Chapter 1 that this is sustained because of feedback asymmetry: there are immediate and acute productive gains, and little or no feedback about any gathering danger, particularly if the procedure was successful, and, in Snook's case, if lack of contact with neighboring units hides the size and nature of the growing gaps between different local adaptations.

Plans and procedures, then, for dealing with potentially safety-critical tasks, are always subject to local modification. Those charged with implementing and following them find ways to work around aspects of plans or protocols that impede fluid accomplishment of their tasks. What can happen is that these local adaptations drift further and further from the original rationale for implementing the tighter procedure or protocol in the first place. This can occur in large part because of structural secrecy. Little may be seen or known, either vertically or horizontally in an organization, about how local groups adapt practice. Drift occurs in the seams between departments, professions, specializations or hierarchies. It slowly but steadily decouples localized realizations of the activity from the original centralized task design. The local deviance from centralized impositions or expectations is slow and sensible enough (and successful enough) to be virtually unnoticeable, particularly from the inside that local work group. It becomes acceptable and expected behavior; the deviant becomes the norm. Risk is not seen, because it is packed away in the gradual acceptance of deviance (and its conversion into normality) within local work groups.

In loosely coupled situations (remember normal accidents theory from above), this may not be a problem. Routine situations, in which there is time to reflect, to stop, to start over, can easily absorb adaptations and improvisations, even if these are not really coordinated across professional or hierarchical boundaries. People or groups may not have to shed their locally evolved rational actions and habits, because it doesn't really have negative consequences. There is always margin, or slack, to recover and repair misunderstanding or confusion.

But in a rapidly deteriorating situation, coupling becomes tighter. Time can be of the essence, things have to be done in a fixed, rapid order, and substitution of material or expertise or protocols is not possible. Snook showed that if various local practices have drifted apart over time (even if that makes local sense), by not coordinating their adaptations or improvisations with others, it may become really difficult to act smoothly and successfully in a tightly coupled situation. When different practitioners or teams come together to solve a problem that is outside the routine, and for which they usually don't see each other, the effects of drift quickly come to the surface. Some people's roles may have subtly shifted, material may have been sorted and placed elsewhere, objects may have begun to be called by different names. Risk gets a chance to express itself only when different locally adapted practices meet. Structural secrecy and practical drift feed on each other: not knowing what other units do allows work groups to drift into locally practical arrangements. An increased distance between how individual groups work fosters further structural secrecy because, at least in loosely coupled situations, they will have little or even less reason or opportunity to communicate with each other.

MANAGING THE INFORMATION ENVIRONMENT

If the seeds of drift into failure lie in the normal, daily decisions and trade-offs that people inside work groups make, and if production pressures and sacrifices to safety have become pre-rational influences in such decision-making, then one place to try to intervene is exactly there. What information, really, are people at all levels in an organization using to base their decisions on? The way in which information is selected and presented of course has consequences for what people will see as the problem to be solved, and what aspects of that problem are relevant and which are not. So how is it presented, by whom? Recall the high-reliability recommendations to insert more expertise into these decision-making processes, and for decision-makers to take minority opinion seriously. Studying and enhancing the "information environment" for decision-making, as Rasmussen and Svedung put it, can be a good place to start.[46] This information environment, after all, is where assessments are made, decisions are shaped, in which local rationality is created. It is the place where the social and the technical meet; where risk itself is constructed.

> The two Space Shuttle accidents (*Challenger* in 1996 and *Columbia* in 2002) are highly instructive here, if anything because the *Columbia* Accident Investigation Board (CAIB), as well as later analyses of the *Challenger* disaster represent significant (and, to date, rather unique) departures from the typical structuralist probes into such accidents. These analyses take normal organizational processes toward drift seriously, applying and even extending a language that helps us capture something essential about the continuous creation of local rationality by organizational decision-makers.[47]

> One critical feature of the information environment in which NASA engineers made decisions about safety and risk was "bullets." Richard Feynman, who participated in the original Rogers Presidential Commission investigating the *Challenger* disaster, already fulminated against them and the way they collapsed engineering judgments into crack statements: "Then we learned about 'bullets' – little black circles in front of phrases that were supposed to summarize things. There was one after another of these little goddamn bullets in our briefing books and on the slides."[48]

"Bullets" appear again as an eerie outcropping in the 2003 *Columbia* accident investigation. With the proliferation of commercial software for making "bulletized" presentations since *Challenger*, bullets proliferated as well. This too may have been the result of locally rational (though largely unreflective) trade-offs to increase efficiency: Bulletized presentations collapse data and conclusions and are dealt with more quickly than technical papers. But bullets filled up the information environment of NASA engineers and managers at the cost of other data and representations. They dominated technical discourse and, to an extent, dictated decision-making, determining what would be considered as sufficient information for the issue at hand. Bulletized presentations were central in creating local rationality, and central in nudging that rationality ever further away from the actual risk brewing just below.

Edward Tufte analysed one *Columbia* slide in particular, from a presentation given to NASA by a contractor in February 2003. The aim of the slide was to help NASA consider the potential damage to heat tiles created by ice debris that had fallen from the main fuel tank. (Damaged heat tiles triggered the destruction of *Columbia* on the way back into the earth's atmosphere.) The slide was used by the Debris Assessment Team in their presentation to the Mission Evaluation Room. It was entitled "Review of Test Data Indicates Conservatism for Tile Penetration," suggesting, in other words, that the damage done to the wing was not so bad.[49] But actually, the title did not refer to predicted tile damage at all. Rather, it pointed to the choice of test models used to predict the damage. A more appropriate title, according to Tufte, would have been "Review of Test Data Indicates Irrelevance of Two Models." The reason was that the piece of ice debris that struck the *Columbia* was estimated to be 640 times larger than the data used to calibrate the model on which engineers based their damage assessments (Later analysis showed that the debris object was actually 400 times larger). So the calibration models were not of much use: They hugely underestimating the actual impact of the debris. The slide went on to say that "significant energy" would be required to have debris from the main tank penetrate the (supposedly harder) tile coating of the Shuttle wing, yet that test results showed that this was possible at sufficient mass and velocity, and that, once the tiles were penetrated, significant damage would be caused.

As Tufte observed, the vaguely quantitative word "significant" or "significantly" was used five times on the one slide, but its meaning ranged all the way from the ability to see it using those irrelevant calibration tests, through a difference of 640-fold, to damage so great that everybody onboard would die. The same word, the same token on a slide, repeated five times, carried five profoundly (yes, significantly) different meanings, yet none of those were really made explicit because of the condensed format of the slide. Similarly, damage to the protective heat tiles was obscured behind one little word, it, in a sentence that read "Test results show that it is possible at sufficient mass and velocity."[50] The slide weakened important material, and the life-threatening nature of the data on it was lost behind bullets and abbreviated statements.

A decade and a half before, Feynman had discovered a similarly ambiguous slide about *Challenger*. In his case, the bullets had declared that the eroding seal in the field joints was "most critical" for flight safety, yet that "analysis of existing data indicates that it is safe to continue flying the existing design."[51] The accident proved that it was not. Solid Rocket Boosters (or SRBs or SRMs) that help the Space Shuttle out of the earth's atmosphere are segmented, which makes ground transportation easier and has some other advantages. A problem that was discovered early in the Shuttle's operation, however, was that the solid rockets did not always properly seal at these segments, and that hot gases could leak through the rubber O-rings in the seal, called blow-by. This eventually led to the explosion of *Challenger* in 1986. The pre-accident slide picked out by Feynman had declared that while the lack of a secondary seal in a joint (of the solid rocket motor) was "most critical," it was still "safe to continue flying." At the same time, efforts needed to be "accelerated" to eliminate SRM seal erosion. During *Columbia* as well as *Challenger*, slides were not just used to support technical and operational decisions that led up to the accidents. Even during both post-accident investigations, slides with bulletized presentations were offered as substitutes for technical analysis and data, causing the *Columbia* Accident Investigation Board, similar to Feynman years before, to conclude that: "The Board views the endemic use of PowerPoint briefing slides instead of technical papers as an illustration of the problematic methods of technical communication at NASA."[52]

The overuse of bullets and slides illustrates the problem of information environments and how studying them can help us understand something about the creation of local rationality in organizational decision-making. NASA's bulletization shows how organizational decision-makers are configured in an impoverished information environment. That which decision-makers can know is generated by other people, and gets distorted during transmission through a reductionist, abbreviated medium. The narrowness and incompleteness of the environment in which decision-makers find themselves can come across as disquieting to retrospective observers, including people inside and outside the organization. It was after the *Columbia* accident that the Mission Management Team "admitted that the analysis used to continue flying was, in a word, 'lousy.' This admission – that the rationale to fly was rubber-stamped – is, to say the least, unsettling."[53]

Unsettling it may be, and probably is – in hindsight. But from the inside, people in organizations do not spend a professional life making "unsettling" decisions. Rather, they do mostly normal work. Again, how can a manager see a "lousy" process to evaluate flight safety as normal, as not something that is worthy reporting or repairing? How could this process be normal? As Vaughan did with

Challenger, the *Columbia* Accident Investigation Board found clues to answers in pressures of scarcity and competition: "The Flight Readiness process is supposed to be shielded from outside influence, and is viewed as both rigorous and systematic. Yet the Shuttle Program is inevitably influenced by external factors, including, in the case of STS-107, schedule demands. Collectively, such factors shape how the Program establishes mission schedules and sets budget priorities, which affects safety oversight, workforce levels, facility maintenance, and contractor workloads. Ultimately, external expectations and pressures impact even data collection, trend analysis, information development, and the reporting and disposition of anomalies. These realities contradict NASA's optimistic belief that pre-flight reviews provide true safeguards against unacceptable hazards."[54]

Studying information environments, how they are created, sustained, and rationalized, and in turn how they help support and rationalize complex and risky decisions, is one route to understanding the small incremental steps that an organization makes towards its margins. Managing the information environment, of course, is not something that can be done with a priori decisions about what is important and what is not. Because that simply displaces the problem to an environment prior to the one that needs to be influenced. Also, a priori knowledge of what is important is very difficult to establish in complex systems with unruly technology. The very nature of these systems, and the technology they operate, make predictions about what is going to fail and when virtually impossible. This is why high reliability theory recommends decision-makers to remain complexly sensitized; to live in an information environment that is full of inputs from all kinds of sides and angles. Yet this can create a signal-to-noise ratio problem for the decision-maker. And it can once again encourage a tendency to oversimplify, to categorize, to bulletize.

Recall that high reliability theory also encourages decision-makers to defer to expertise and to take minority opinion seriously. This should enrich their information environment too. But even this does not necessarily help. As indicated above, perhaps there is no such thing as "rigorous and systematic" decision-making based on technical expertise alone. Expectations and pressures, budget priorities and schedules, contractor workloads, employee qualifications and workforce levels all impact technical decision-making. All these factors determine and constrain what will be seen as possible and rational courses of action at the time, even by experts. Although the intention was that NASA's flight safety evaluations be shielded from external pressures (turning it into a model closed system, as per the high-reliability recommendation), these pressures nonetheless seeped into even the collection of data, analysis of trends and reporting of anomalies. The information environments thus created for decision-makers were continuously and insidiously tainted by pressures of production and scarcity (and in which organization are they not?), pre-rationally influencing the way people saw the world. Yet even this "lousy" process was considered "normal" – normal or inevitable enough, in any case, to not warrant expending energy and political capital on trying to change it. Drift into failure was the result.

CONTROL THEORY AND DRIFT

A family of ideas that approaches the problem of drift from another angle than the social-organizational one is control theory. Control theory looks at adverse events as emerging from interactions among system components. It usually does not identify single causal factors, but rather looks at what may have gone wrong with the system's operation or organization of the hazardous technology that allowed an accident to take place. Safety, or risk management, is viewed as a control problem, and adverse events happen when component failures, external disruptions or interactions between layers and components are not adequately handled; when safety constraints that should have applied to the design and operation of the technology have loosened, or become badly monitored, managed, controlled. Control theory tries to capture these imperfect processes, which involve people, societal and organizational structures, engineering activities, and physical parts. It sees the complex interactions between those as eventually resulting in an accident.

Control theory sees the operation of hazardous processes as a matter of keeping many interrelated components in a state of dynamic equilibrium. This means that control inputs, even if small, are continually necessary for the system to stay safe: like a bicycle, it cannot be left on its own, or it would lose balance and collapse. A dynamically stable system is kept in equilibrium through the use of feedback loops of information and control. Adverse events are not seen as the result of an initiating event or root cause that triggers a linear series of events. Instead, adverse events result from interactions among components that violate the safety constraints on system design and operation. Feedback and control inputs can grow increasingly at odds with the real problem or processes to be controlled. Concern with those control processes (how they evolve, adapt and erode) lies at the heart of control theory as applied to organizational safety.

Control theory says that the potential for failure builds because deviations from the system's original design assumptions become increasingly rationalized and accepted. This is consistent with Vaughan's and Snook's descriptions of the social systems in which decision-making occurs and local actions and rationalities develop. Adaptations occur, adjustments get made, and constraints get loosened in response to local concerns with limited time-horizons. They are all based on uncertain, incomplete knowledge. Just like practical drift and structural secrecy, this can engender and sustain erroneous expectations of users or system components about the behavior of others in the system.

A changed or degraded control structure eventually leads to adverse events. In control-theoretic terms, degradation of the safety-control structure over time can be due to asynchronous evolution, where one part of a system changes without the related necessary changes in other parts. Changes to subsystems may have been carefully planned and executed in isolation, but consideration of their effects on other parts of the system, including the role they play in overall safety control, may remain neglected or inadequate. Asynchronous evolution can occur, too, when one part of a properly designed system deteriorates independent of other parts.

The more complex a system (and, by extension, the more complex its control structure), the more difficult it can become to map out the reverberations of changes (even carefully considered ones) throughout the rest of the system. Control theory embraces a more complex idea of causation than the energy-to-be-contained models discussed above (see also Chapter 8). Small changes somewhere in the system, or small variations in the initial state of a process, can lead to large consequences elsewhere.

Control theory helps in the design control and safety systems (particularly software-based) for hazardous industrial or other processes.[55] When applied to organizational safety, control theory is concerned with how an erosion of a control structure allows a migration of organizational activities towards the boundary of acceptable safety performance.

Leveson and her colleagues applied control theory to the analysis of a water contamination incident that occurred in May 2000 in the town of Walkerton, Ontario, Canada.[56] The contaminants E. coli and Campylobacter entered the water system through a well of the Walkerton municipality, which operated the system through its Walkerton Public Utilities Commission (WPUC). Leveson's control theoretic approach showed how the incident flowed from a steady (and rationalized, normalized) erosion of the control structure that had been put in place to guarantee water quality.

The proximate events were as follows. In May 2000, the water system was supplied by three groundwater sources: wells 5, 6, and 7. The water pumped from each well was treated with chlorine before entering the distribution system. The source of the contamination was manure that had been spread on a farm near well 5. Unusually heavy rains from May 8 to May 12 carried the bacteria to the well. Between May 13 and 15, a WPUC employee checked well 5 but did not take measurements of chlorine residuals, although daily checks were supposed to be made. Well 5 was turned off on May 15 and well 7 was turned on. A new chlorinator, however, had not been installed on well 7 and the well was therefore pumping unchlorinated water directly into the distribution system. The WPUC employee did not turn off the well, but instead allowed it to operate without chlorination until noon on Friday May 19, when the new chlorinator was installed.

On May 15, samples from the Walkerton water distribution system were sent to a laboratory for testing according to the normal procedure. Two days later, the laboratory advised WPUC that samples from May 15 tested positive for E. coli and other bacteria. On May 18, the first symptoms of illness appeared in the community. Public inquiries about the water prompted assurances by WPUC that the water was safe. The next day, the outbreak had grown, and a physician contacted the local health unit with a suspicion that she was seeing patients with symptoms of E. coli.

In response to the lab results, WPUC started to flush and superchlorinate the system to try to destroy any contaminants in the water. The chlorine residuals began to recover. WPUC did not disclose the lab results. They continued to flush and superchlorinate the water through the following weekend, successfully increasing the chlorine residuals. Ironically, it was not the operation of well 7 without a chlorinator that caused the contamination; the contamination instead entered the system through well 5 from May 12 until it had been shut down on May 15.

Without waiting for more samples, the community issued a boil water advisory on May 21. About half of Walkerton's residents became aware of the advisory on May 21, with some members of the public still drinking the Walkerton town water as late as May 23. Seven people died and more than 2,300 become ill.

The proximate events could be modeled using a sequence-of-events approach, which would point to the various errors and violations and shortcomings in the systems layers of defense. But Leveson and colleagues decided to model the Ontario water quality safety control structure and show how it eroded over time, allowing the contamination to take place. The safety control structure was intended to prevent exposure of the public to contaminated water, first by removing contaminants, second by public health measures that would prevent consumption of contaminated water (see Figure 5.1).

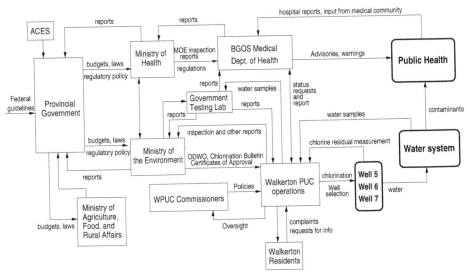

Figure 5.1   Control structure as originally envisioned to guarantee water quality in Walkerton (Leveson, Daouk et al. 2003)

In Ontario, decisions had been taken to remove various water safety controls, or to reduce their enforcement, without an assessment of the risks. One of the important features that disappeared were feedback loops. As the other controls weakened or disappeared over time, the entire socio-technical system moved to a state where a small change in the operation of the system or in the environment (in this case, unusually heavy rain) could lead to a tragedy.

Well 5 had been vulnerable to contamination to begin with. It was shallow, in an area open to farm runoff, and perched on top of bedrock with only a thin layer of top soil around it. No extra approval for the well had been necessary, however, and it was connected to the municipal system as a matter of routine. No program or policy was in place to review existing wells to determine whether they met requirements or needed continuous monitoring.

A number of factors led to erosion of the control structure. These included objections to the taste of chlorine in drinking water, WPUC employees who could safely consume untreated water from the wells, a lack of certification for water system operators, inexperience with water quality processes, and a focus on financial strains on WPUC. A lack of government policy on land use and watershed exposed this increasingly brittle structure to heavily contaminated water by hog and cattle farming. Budget and staff reductions by a new

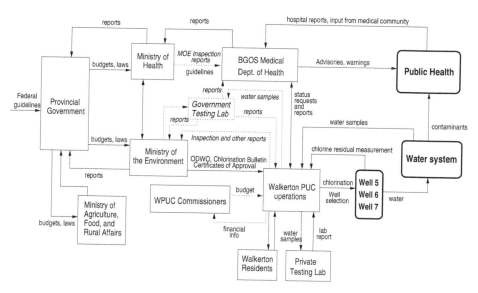

Figure 5.2 **Safety control structure at the time of the water contamination incident at Walkerton. Controls had been loosened, feedback loops had disappeared. The control structure had become hollowed-out relative to its original design intentions (Leveson, Daouk et al. 2003)**

conservative government took a toll on environmental programs and agencies. A Water Sewage Services Improvement Act was passed in 1996, which shut down the government-run testing laboratories, delegated control of provincially owned water and sewage plants to municipalities, eliminated funding for municipal water utilities, and ended the provincial drinking water surveillance program. Farm operators were from now on to be treated with understanding if they were found in violation of livestock and waste-water regulations. No criteria were established to ensure the quality of testing or the qualifications or experience of private lab personnel, and no provisions were made for licensing, inspection, or auditing of private labs by the government. The resulting control structure was a hollowed-out version of its former self. It had become brittle, and vulnerable to an unusual perturbation (like massive rainfall), lacking the resilience or redundancies to stop the problem or recover quickly from it.

Control theory does not see an organization as a static design of components or layers. It readily accepts that a system is more than the sum of its constituent elements. Instead, it sees an organization as a set of constantly changing and adaptive processes focused on achieving the organization's multiple goals and adapting around its multiple constraints. The relevant units of analysis in control theory are therefore not components or their breakage (for example, holes in layers of defense), but system constraints and objectives.[57]

An important consequence is that control theory is not concerned with individual unsafe acts or errors, or even individual events that may have helped trigger an adverse event. Such a focus does not help, after all, in identifying broader ways to protect the system against migrations towards risk. Control theory also rejects the depiction of adverse events in a traditionally physical way as the latent failure model does, for example. Accidents are not about particles, trajectories or collisions between hazards and the process-to-be-protected. Removing individual unsafe acts, errors or singular events from an adverse event sequence only creates more space for new ones to appear if the same kinds of systemic constraints and objectives are left similarly ill-controlled. The focus of control theory is therefore not on erroneous actions or violations, but on the mechanisms that help generate such behaviors at a higher level of functional abstraction – mechanisms that turn these behaviors into normal, acceptable and even indispensable aspects of an actual, dynamic, daily work context that needs to survive inside the constraints of three kinds of boundaries (functional, economic and safety).

For control theory, the making and enforcing of rules is not an effective strategy for controlling behavior. This, instead, can be achieved by making the boundaries of system performance explicit and known, and to help people develop skills at coping with the edges of those boundaries.[58] This, indeed, should be part of the information environment in which decision-makers operate. Ways proposed by Rasmussen include increasing the margin from normal operation to the safety boundary. This can be done by moving the safety boundary further out, or by moving operations further inward, away from a fixed safety boundary. In both cases more margin opens up. This, however, is only partially effective because of risk homeostasis – the tendency for a system to gravitate back to a certain level of risk acceptance, even after interventions to make it safer. In other words, if the boundary of safe operations is moved further away, then normal operations will likely follow not long after – under pressure, as they always are, from the objectives of efficiency and less effort.

Leaving both pressures in place (a push for greater efficiency and a safety campaign pressing in the opposite direction) does little to help operational people (pilots in the case above) cope with the actual dilemma at the boundary. Also, a reminder to try harder and watch out better, particularly during times of high workload, is a poor substitute for actually developing skills to cope at the boundary. Raising awareness, however, can be meaningful in the absence of other possibilities for safety intervention, even if the effects of such campaigns tend to wear off quickly. Greater safety returns can be expected only if something more fundamental changes in the behavior-shaping conditions or the particular process environment. In this sense, it is important to raise awareness about the migration toward boundaries throughout the organization, at various managerial levels. To the extent possible, of course. But if such things can somehow be made visible, a fuller range of countermeasures becomes available beyond telling front-line operators to be more careful.

RESILIENCE ENGINEERING

Pressures of scarcity and competition, the intransparency and size of complex systems, the patterns of information that surround decision-makers, and the decrementalist nature of their decisions over time, can cause systems to drift into failure. Drifting into failure poses a substantial risk even for safe systems. Safety is an emergent property, and its erosion is not about the breakage or lack of quality of single components. Drifting into failure is not so much about breakdowns or malfunctioning of components, as it is about an organization not adapting effectively to cope with the complexity of its own structure and environment.

Drift is generated by normal processes of reconciling differential pressures on an organization (efficiency, capacity utilization, safety) against a background of uncertain technology and imperfect knowledge. Drift is about decrementalism contributing to extraordinary events, about the transformation of pressures of scarcity and competition into organizational mandates, and about the normalization of signals of danger so that organizational goals and supposedly normal assessments and decisions become aligned. In safe systems, the very processes that normally guarantee safety and generate organizational success, can also be responsible for organizational demise. The same complex, intertwined socio-technical life that surrounds the operation of successful technology, is to a large extent responsible for its potential failure. Because these processes are normal, because they are part and parcel of normal, functional organizational life, they are difficult to identify and disentangle. The role of these invisible and unacknowledged forces can be frightening. Harmful consequences can occur in organizations constructed to prevent them. Harmful consequences can occur even when everybody follows the rules.

Recall that one of the recent recommendations of high-reliability theory is a commitment to resilience.[59] Ideas about resilience and resilience engineering predate this recommendation,[60] and in fact, when it comes to resilience, there is something unfortunate about the label originally chosen to denote the high-reliability school of thought. Safety is not the same as reliability. A part can be reliable, but in and of itself it can't be safe. It can perform its stated function to some expected level or amount, but it is context, the context of other parts, of the dynamics and the interactions and cross-adaptations between parts, that make a system safe or unsafe. That is, indeed, the whole point of complexity thinking: safety inheres in the hugely complex and ever-changing relationships between parts, not in the parts themselves. Reliability in itself cannot capture this at all. As an engineering property, reliability is expressed as a component's failure rate over a period of time. High reliability means that it will take a long time before the part fails. In other words, reliability addresses the question of whether a component lives up to its pre-specified performance criteria. This, indeed, is what quality improvement and control can be about too. Quality and reliability are often associated with a reduction in variability, and an increase in replicability: the same process, narrowly guarded, produces the same predictable outcomes. But predictability and replicability are not necessarily the features that make a complex system safe.

The direction in which drift pushes the operation of the technology can be hard to detect, also or perhaps especially for those on the inside. It can be even harder to stop. Given the diversity of forces (political, financial, and economic pressures, technical uncertainty, incomplete knowledge, fragmented problem-solving processes) both on the inside and outside, the large, complex sociotechnical systems that operate some of our most hazardous technologies today seem capable of generating an obscure energy and drift of their own, relatively impervious to outside inspection or inside control. Resilience engineering accepts that all open systems are continually adrift inside their safety envelopes. In fact, it takes this premise as a source of organizational strength, rather than a weakness. Organizational resilience is about finding the political, practical and operational means to invest in safety even under pressures of scarcity and competition, because that may be when such investments are needed most. Organizational resilience is seen not as a property, but as a capability: A capability to recognize the boundaries of safe operations, a capability to steer back from them in a controlled manner, a capability to recover from a loss of control if it does occur. And a capability to detect and recognize the phases of operation when the margins are skirted or crossed. An important question studied by resilience engineering is how an organization can monitor and keep track of its own adaptations (and how these bound the rationality of decision-makers) to pressures of scarcity and competition, while dealing with imperfect knowledge and unruly technology. How can an organization become aware, and remain aware, of its models of risk and danger? Preventing a drift into failure requires a different kind of organizational monitoring and learning. It means fixing on higher order variables, adding a new level of intelligence, adopting a new language. The next two chapters will use the language of complexity and systems thinking to initiate just that.

REFERENCES

1   Turner, B.A. (1978). *Man-made disasters*. London: Wykeham.
2   In 1995, Barry Turner wrote "…with *Man-made Disasters*, I had the sense after publication that the book had disappeared without a trace." (Turner, B.A. (1995). A personal trajectory through organization studies. *Research in the Sociology of Organizations*, 13, 275–301, p. 295). Indeed, the book was seldom cited in its first decades of existence. Contemporary scholars find this hard to explain, given the quality of the work and the unprecedented approach he took. The lack of social concern about the threat of technical hazard may have been one factor, though in the U.S. the Three Mile Island nuclear power plant accident happened only a year after the publication of *Man-made disasters*. Perhaps the broader scholarly and societal discussion after that accident focused more on structural systems and engineering aspects rather than administrative and organizational issues, leading, among other things, to a book that did become a classic in the U.S.A. and elsewhere: Perrow's *Normal Accidents* which was first published in 1984.
3   Turner, B.A. and Pidgeon, N.F. (1997). *Man-made Disasters* (Second Edition). Oxford: Butterworth Heinemann, p. 72.
4   Turner, B.A. and Pidgeon, N.F. (1997). Ibid., p. 72.
5   This is the problem of converting a fundamental surprise into a local one: instead of overhauling the beliefs in the safety of the system and the ways to ensure that safety, the disaster will be seen as a local glitch, due to some technical component failure or human error. This is much easier, cheaper and more convenient, of course, than reconstructing beliefs about safety and risk (and having to make concomitant system investments).
6   Stech, F.J. (1979). *Political and Military Intention Estimation. Report N00014–78–0727.* Bethesda, MD: US Office of Naval Research, Mathtech Inc.
7   Lanir, Z. (1986). *Fundamental surprise*. Eugene, OR: Decision Research.
8   Turner, B.A. and Pidgeon, N.F. (2000). Man-made disasters: Why technology and organizations (sometimes) fail. *Safety Science*, 34, 15–30.
9   Weick, K.E. and Sutcliffe, K.M. (2007). *Managing the unexpected: Resilient performance in an age of uncertainty*. San Fracisco: Jossey-Bass, p. 93.
10   Reason, J.T. (1990). Human error. Cambridge, UK: Cambridge University Press.
11   Wijaya, M. (2002). *Architecture of Bali: A source book of traditional and modern forms*. Honolulu: University of Hawai'i Press.
12   Turner, B.A. and Pidgeon, N.F. (2000). Ibid., p. 16.
13   Marone, J.G. and Woodhouse, E.J. (1986). *Avoiding catastrophe: Strategies for regulating risky technologies*. Berkeley, CA: University of California Press.
14   Rochlin, G.I., LaPorte, T.R. et al. (1987). The self-designing high reliability organization: Aircraft carrier flight operations at sea. *Naval War College Review*, 76–90.

15   Sagan, S.D. (1993). *The limits of safety: Organizations, accidents and nuclear weapons.* Princeton, NJ: Princeton University Press.

16   Rochlin, G.I., LaPorte, T.R. et al. (1987). Ibid.

17   LaPorte, T.R. and Consolini, P.M. (1991). Working in Practice but Not in Theory: Theoretical Challenges of 'High-Reliability Organizations.' *Journal of Public Administration Research and Theory: J-PART 1*(1): 19–48.

18   Wildavsky, A.B. (1988). *Searching for safety.* New Brunswick: Transaction Books.

19   Rochlin, G.I. (1999). Safe operation as a social construct. *Ergonomics*, 42(11), 1549–60, p. 1549.

20   Weick, K.E. (1993). The collapse of sensemaking in organizations: The Mann-gulch disaster. *Administrative Science Quarterly*, 38(4), 628–52.

21   Weick, K.E. and Sutcliffe, K.M. (2007). Ibid.

22   Starbuck, W.H. and Milliken, F.J. (1988). Challenger: Fine-Tuning the Odds Until Something Breaks. *The Journal of Management Studies*, 25(4), 319–41, pp. 329–30.

23   Weick, K.E. and Sutcliffe, K.M. (2007). Ibid., p. 52.

24   Weick, K.E. and Sutcliffe, K.M. (2007). Ibid.

25   Jensen, C. (1996). *No downlink: A dramatic narrative about the Challenger accident and our time.* New York: Farrar, Strauss, Giroux, p. 368.

26   Perrow, C. (1984). *Normal accidents: Living with high-risk technologies.* New York: Basic Books, p. 23.

27   Weick, K.E. (1995). Sensemaking in organizations. London: Sage, p. 87.

28   See Dekker, S.W.A. (2005). *Ten questions about human error: A new view of human factors and system safety.* Mahwah, NJ: Lawrence Erlbaum Associates.

29   Dörner, D. (1989). *The logic of failure: Recognizing and avoiding error in complex situations.* Cambridge, MA: Perseus Books, p. 65.

30   National Aeronautics and Space Administration. (2000, March). *Report on project management in NASA, by the Mars Climate Orbiter Mishap Investigation Board.* Washington, DC: Author. See also Dekker, S.W.A. (2005). *Ten questions about human error: A new view of human factors and system safety.* Mahwah, NJ: Lawrence Erlbaum Associates.

31   Starbuck, W.H. and Milliken, F.J. (1988). Ibid., p. 323.

32   *Airliner World* (2001, November), p. 79.

33   Dörner, D. (1989). *The logic of failure: Recognizing and avoiding error in complex situations.* Cambridge, MA: Perseus Books, p. 68.

34   See Dekker, S.W.A. (2005). Ibid.

35   McDonald, N., Corrigan, S., and Ward, M. (2002, June). *Well-intentioned people in dysfunctional systems.* Keynote paper presented at the 5th Workshop on Human error, Safety and Systems development, Newcastle, Australia.

36   McDonald, N., Corrigan, S., and Ward, M. (2002, June). Ibid.

37   Vaughan, D. (1996). The Challenger launch decision: Risky technology, culture and deviance at NASA. Chicago: University of Chicago Press.

38   Snook, S.A. (2000). *Friendly fire: The accidental shootdown of US Black Hawks over Northern Iraq.* Princeton, NJ: Princeton University Press.

39   Starbuck, W.H. and F.J. Milliken (1988). Ibid., p. 333.

40   Vaughan, D. (1996). Ibid., p. xiv.

41   Vaughan, D. (1996). Ibid., p. 420.

42 Vaughan, D. (1996). Ibid., p. 416.

43 Cohen, M.D., J.G. March, et al. (1988). A garbage can model of organizational choice. In J.G. March (ed.), *Decisions and organizations*, pp. 294–334. Oxford: Blackwell.

44 Vaughan, D. (1996). Ibid., p. 250.

45 Snook, S.A. (2000). *Friendly fire: The accidental shootdown of US Black Hawks over Northern Iraq.* Princeton, NJ: Princeton University Press, p. 182.

46 Rasmussen, J., and Svedung, I. (2000). *Proactive risk management in a dynamic society.* Karlstad, Sweden: Swedish Rescue Services Agency.

47 See Dekker, S.W.A. (2005). Ibid.

48 Feynman, R.P. (1988). *"What do you care what other people think?": Further adventures of a curious character.* New York: Norton.

49 Columbia Accident Investigation Board. (2003, August). *Report Volume 1.* Washington, DC: CAIB, p. 191.

50 Columbia Accident Investigation Board. (2003, August). Ibid., p. 191.

51 Feynman, R.P. (1988). Ibid., p. 137.

52 Columbia Accident Investigation Board. (2003, August). Ibid., p. 191.

53 Columbia Accident Investigation Board. (2003, August). Ibid., p. 190.

54 Columbia Accident Investigation Board. (2003, August). Ibid., p. 191.

55 Leveson, N.G. and Turner, C.S. (1993). An investigation of the Therac-25 accidents. *Computer*, 26(7), 18–41.

56 Leveson, N., Daouk, M., et al. (2003). *Applying STAMP in accident analysis.* Cambridge, MA, Engineering Systems Division, Massachusetts Institute of Technology.

57 Leveson, N.G. (2002). *A new approach to system safety engineering.* Cambridge, MA, Aeronautics and Astronautics, Massachusetts Institute of Technology.

58 Rasmussen, J. (1997). Risk management in a dynamic society: A modeling problem. *Safety Science*, 27(2–3), 183–213.

59 Weick, K.E. and Sutcliffe, K.M. (2007). Ibid.

60 Hollnagel, E., Woods, D.D., and Leveson, N.G. (eds.) (2006). Resilience engineering: Concepts and precepts. Aldershot, UK: Ashgate Publishing Co.

6

WHAT IS COMPLEXITY AND SYSTEMS THINKING?

In trying to understand why things go wrong in complex systems, the safety specialist has traditionally looked at the functioning or breakdown of carefully isolated parts or phenomena. How is it that the pilot did not notice a speed decay? How can it be that the nurse did not see the bold-printed warning EPIDURAL ONLY on the infusion bag? Why did a piece of insulating foam separate from the left bipod ramp section of the Space Shuttle *Columbia* external tank at 81.7 seconds into the launch?

These are important questions, and often hold the key to unlocking at least part of a story on how things went wrong. These questions make it possible to research and systematize various detailed processes, or to find out about the state of a part or a component at or before the time of some failure. But there is something really important that such knowledge cannot tell us, and that is how a number of different things and processes act together when exposed to a number of different influences at the same time.[1] This is the question that complexity and systems thinking tries to answer.

MORE REDUNDANCY AND BARRIERS, MORE COMPLEXITY

The Newtonian focus on parts as the cause of accidents has sponsored a belief that redundancy is the best way to protect against hazard. Safety-critical systems usually have multiple redundant mechanisms, safety systems, elaborate policies and procedures, command and reporting structures as well as professional specialization. The results of such organizational, operational and engineering measures makes these systems relatively safe from single point failures. It protects them against the failure of a single component or part that could directly lead to a bad outcome. But there is a cost to this.

Barriers, as well as professional specialization, policies, procedures, protocols, redundant mechanisms and structures add to a system's complexity. With the introduction of each new part or layer of defense, there is an explosion of new relationships (between parts and layers and components) that spreads through the system. Complexity and systems thinking says that accidents emerge from these relationships, even from perfectly "normal" relationships, where nothing (not even a part) is broken. The drive to make these systems reliable, in other words, also makes them very complex. Which, paradoxically, can make them less safe. Redundancy, or putting in extra barriers, or fixing them up better, does not provide protection against this safety threat. In fact, it helps create it, by making the systems more interactively complex.[2]

Interactive complexity refers to component interactions that are non-linear, unfamiliar, unexpected or unplanned, and either not visible or not immediately comprehensible for people running the system. Linear interactions among components, in contrast, are those in expected and familiar production or maintenance sequences, and those that are visible and understandable even if they were unplanned. But complex interactions produce unfamiliar sequences, or unplanned and unexpected sequences, that are either not visible or not immediately comprehensible. An electrical power grid is an example of an interactively complex system. Failures, when they do occur, can cascade through these systems in ways that may confound the people managing them, making it difficult to stop a progression of failure. Increasing complexity, particularly interactive complexity, has thus given rise to a new kind of accident over the last half century: the system accident.[3] In a system accident, no component needs to be broken. The accident results from the relationships between components (or the software and people running them), not from the workings or dysfunction of any component part.

This insight has grown out of what became known as systems engineering and system safety, pioneered in part by 1950s aerospace engineers who were confronted with increasing complexity in aircraft and ballistic missile systems at that time. One particular aspect of complexity fascinated and concerned safety researchers: the interconnectivity and interactivity between system components. Their relationships, in other words. The problem with this was that greater complexity led to vastly more possible interactions than could be planned, understood, anticipated or guarded against. Rather than being the result of a few or a number of component failures, accidents involve the unanticipated interaction of a multitude of events in a complex system – events and interactions, often very normal, whose combinatorial explosion can quickly outwit people's best efforts at predicting and mitigating trouble. Interactive complexity refers to component interactions that are non-linear, unfamiliar, unexpected or unplanned, and either not visible or not immediately comprehensible for people running the system.[4] The idea behind system accidents is that our ability to intellectually manage interactively complex systems has now been overtaken by our ability to build them and let them grow (like a NASA-contractor bureaucratic-organizational complex).

Just think of the introduction a new procedure to double-check something that was implicated as a broken part in some previous accident. The new procedure relates to the old procedure and its remnants in people's memories and rehearsed action sequences. It relates to how people need to carry it out in context, to people who have to train the new procedure, to the regulator who may need to approve it. It may take time and attention away from other tasks, which in turn can create a host of reverberations throughout the system.

> "Closed loop communication" was introduced as a safety practice on ships to avoid misunderstandings. Closed loop means that both the receipt and one's understanding of a message are explicitly acknowledged, for example by repeating the message. A maritime pilot, however, had observed that under certain conditions in his pilotage area, the changes of course and speed came so fast, that the helmsman's repeating of the pilot's instructions resulted in them talking at the same time (producing overlapping speech). This resulted in more communication, with more overlap. So he abandoned closed loop communication.[5]

The introduction of automation as a layer of redundancy in many systems has produced the same sorts of ironies. It has created new work, a kind of work that people generally are not good at (monitoring very reliable machinery), and introduced a host of new pathways to breakdown.[6] In fact, the explosive growth of software has added greatly to systems' interactive complexity. With software, the possible states that a system can end up in become mind-boggling. Nancy Leveson and her colleagues at MIT estimated that TCAS II, the second-generation airborne collision prediction and avoidance system now onboard all large airplanes (with more than a million lines of software code), could generate 1,040 states.[7]

How software and redundancy don't really fit together in the classical mechanical sense is apparent from a NASA study of an experimental aircraft that was executing two versions of the control system. It found that all the software problems occurring during flight testing resulted from errors in the redundancy management system and not in the much simpler control software itself, which worked perfectly.[8] Also, software removes many of the natural constraints that held engineers back in introducing even greater complexity to their systems.[9] It doesn't weigh anything, it is easy to modify, and the only capacity constraints on its functioning are system memory and processor speed, two things that have grown explosively too.

> An example of a software-based system that led to multiple deaths without any broken components is the Therac-25 radiation therapy machine (technically called medical linear accelerators (linacs). These accelerate electrons to create high-energy beams that can destroy tumors with minimal impact on the surrounding healthy tissue).[10] Radiation-treatment machines were in enormous demand in hospitals throughout North America, and the Therac equipment, principally controlled by software, was widely considered the best in a growing field.

After a number of suspicious cases of overdosing by Therac-25 machines, a physicist and technician became fascinated by a "Malfunction 54" message that had flashed on the screen during a treatment (the patient fell into a coma and died three weeks later). They typed and retyped the prescription into the computer console, determined to re-create Malfunction 54. They went to the bottom of the screen and then moved the cursor up to change the treatment mode from x-ray mode to electron mode, over and over, for hours. Finally they got the malfunction.

What made the difference was the speed with which the instructions were entered. The computer would not accept new information on a particular phase of treatment (in this case changing the x-ray mode to electron mode) if the technician made the changes within eight seconds after reaching the end of the prescription data. That's what Malfunction 54 meant (but didn't say). If the changes were made so soon, all the new screen data would look correct to the technician, but inside the computer, the software would already have encoded the old information.

That meant the beam on the Therac-25 was set for the much stronger dose needed for an x-ray beam while the turntable was in the electron position. The software running inside the computer included no check to verify that various parts of the prescription data agreed with one another.[11]

Accidents in software-rich systems, such as the Therac-25 but also the Ariane 5 rocket or the NASA Mars Polar Lander, show that none of the components in these systems had to fail in order for things to go terribly wrong. All components can meet their specified design requirements, and still a failure can occur.[12] Better component design, or better protection against failures of individual components (for example, barriers, redundancy) is not the answer to such failures. It is not the answer to systems accidents. The answer lies in understanding relationships. Which is what complexity and systems thinking try to do, particularly when applied to socio-technical organizations.

UP AND OUT, NOT DOWN AND IN

Not long ago, a colleague and I were asked to help a maintenance company find out what factors had "caused" a recent increase in safety events and incidents. For a few months we held our ear to the ground – walked around, interviewed people throughout the organization, even sent out a survey which people answered with enthusiasm. Then the day came to report back to management. "So," my colleague asked me, "what do we say the causes are?"

I was quiet for a moment. Because what could we say in response to that question? I had no idea what the "causes" were, even after doing all the analytic work together. It was fascinating, in fact, that management wanted the "causes" of their problems pinpointed, just like in a broken machine. They wanted us to tell them which screws to tighten, which lines to replace, which areas to lubricate. This was not surprising, probably, for a company whose business consists of doing just that.

But such a Newtonian–Cartesian, or mechanistic understanding of organizations is not limited to people who are in the mechanics business. My undergraduate professor in industrial psychology, decades ago, illustrated the same thinking by putting his hands beside his ear, as if he were holding a small box in the air. "What you want to do is shake that organization around, to find out what's loose in there, what's rattling," he said. That was the recommendation to a large group of wannabe consultants and organizational psychologists. Shake the box, open the box, find the loose part, fix it, close the box, hand it back.

In the case of the maintenance organization that my colleague and I tried to help more recently, we never found what was rattling loose inside. We did identify various broad influences that seemed to shape how people related to their work, what they found important, what they felt encouraged to pay attention to. We began to trace how a number of different things and processes acted together when exposed to a number of different influences at the same time.

The recent spin-off of the maintenance organization, that aimed to have it stand on its own commercial legs, had led to a tension between intensive and extensive growth – should the company nurse the depth and quality of its core processes for its only large customer (the organization of which it had, until recently, been part?) or should it aggressively court outside business and broaden its offerings? Organizations like this produce means-ends oriented social action through its formal structures and processes. Those aim to achieve certainty, conformity and goal attainment. But if you suddenly expose the organization to a new competitive environment, which offers a host of previously unencountered and unexplored threats and opportunities, it can quickly wash away the basis of certainty for what goals need attaining, for what social expectations there are and which ones are considered legitimate by whom, for what norms should be conformed with.

People in core business units stopped reporting incidents: aspects of the new environment made the entire notion of incident negotiable, not just for those facing them on the work floor, but also for those hearing about them higher up. Even when things were reported, little to nothing was done. How exactly such larger influences of a new competitive environment travel down to the heads of individual employees to help determine what they will see as rational at any moment in their work is still anybody's guess, even that of current theorizing. So we could never "pinpoint" the organizational spin-off as a cause.

To make things easy, our story could have ended with a description of the limits on cognitive performance by individuals (the "parts" in the system), how fatigue, language difficulties in the communication with contractors, lack of motivation, goal displacement and role ambiguity all constrained the ability of individuals to do their jobs conscientiously, thus leading to incidents.

But we tried to lift the story out of a fixation with parts. We refused to just go down and into the box to find what was rattling. We went up and out, trying to sketch the influences of the new environment on the organization's structure and relationships (management, individuals, complexity), its processes (patterns of interaction that produce action, including deviance and mistake) and people's tasks (their technical uncertainty and inexactitude, the imperfect knowledge and interpretive flexibility necessary to go from written guidance to action in context). We never enumerated certainties, only possibilities. We never pinpointed, only sketched. That is what complexity and systems thinking does.

Mechanistic thinking about failures, that is, the Newtonian–Cartesian approach, means going down and in. Understanding why things went wrong comes from breaking open the system, diving down, finding the parts, and identifying which ones were broken. This approach is taken even if the parts are located in different areas of the system, such as procedural control, supervisory layers, managerial levels, regulatory oversight. In contrast, systems thinking about failures means going up and out. Understanding comes from seeing how the system is configured in a larger network of other systems, of tracing the relationships with those, and how those spread out to affect, and be affected by, factors that lie far away in time and space from the moment things went wrong.

Take a simple question: How does a cell phone work? To understand systems thinking, or the difference between going down and in versus up and out, this is actually a useful question. The mechanistic answer would begin by screwing the back plate off the phone. What are the parts that are inside? How do they get energized, how do they interact, how do electrical signals flow from input to output? And then go deeper still. What is the software that governs these flows, where are the bits and bytes that hold it all up? The behavior of the phone can be brought down to the behavior of its individual components. It is just the aggregate of all of those component behaviors.

The systemic answer would do something radically different. It would go up and out, locating the cell phone in the complex ecological web of markets and the flows of goods and people that supply its parts, and create the demand for it. One place where the answer could end up is the massive Kahuzi-Biega National Park in eastern Congo in Africa, where the elephant population has been virtually wiped out, and a previously fairly healthy gorilla population has been reduced 50 percent to 10 percent of its original size.[13]

The prize to be found there is Coltan, short for Columbite-Tantalite, a metallic ore. When refined, Coltan becomes a heat resistant powder with unique properties for storing electrical charge (tantalum capacitors control current flow in cell phone circuit boards). Eighty percent of it, worldwide, comes from the eastern Congo. Coltan is manually mined by groups of men. They don't need much more than a shovel. They dig basins, or craters, in streams by scraping off the surface mud. They then slosh the water around the crater, which causes the Coltan ore to settle to the bottom where it is then retrieved by the miners. A group can harvest one kilogram (about two pounds) of Coltan per day. When economic times are good, and cell phones are in demand, a Coltan miner can earn as much as US$200 per month, compared to a salary of US$10 per month for the average Congolese worker.

The cell phone, through the mining of Coltan, links with local civil and cross-national wars. In less than 18 months, the Rwandan army made a quarter of a billion dollars from selling Coltan. The strange thing is that Rwanda has no Coltan. But it borders on the eastern Congo. From there, links go up to Brussels and Antwerp. Vendors in Belgium, the colonial master of Zaire (Congo's predecessor) play a role in smuggling and reselling Coltan. The cell phone also links to gorillas and elephants. In areas where Coltan is mined, the ground is cleared to make mining easier, destroying and contaminating the animals' habitat. The cell phone also links to poverty, exploitation and child labor, which in turn has its basis in the displacement of the local populations by the miners and roving armies. Which again links back to the gorillas. Gorillas are killed and their meat is sold as bush meat to the miners and the varying and fluctuating rebel armies that control the area. Alternative sources of Coltan are difficult to find on this planet, and assurances of it being ecologically responsible, humanitarian or "gorilla-safe" are hard to substantiate.

So how does a cell phone work? The systemic answer says little about how a cell phone works as a self-contained, engineered piece of micro-technology. But it could argue that a cell phone works in part by destroying livelihoods while briefly benefiting a few others, that it works by fueling and sustaining civil and regional wars, and by wiping out local populations of elephants and gorillas. That's not a value statement or moral indictment. It is systems thinking: going up and out, trying to understand how influences from many things relate to and impact on many other things; how the system in question (the cell phone) is configured in networks of other systems – social, natural, political and post-colonial, humanitarian, ecological.

SYSTEMS THINKING

The disenchantment with going down-and-in to explain the whole became acute at the turn of the twentieth century. Ironically, it was the same time that the success of the Newtonian model was being celebrated, and its continuing dominance was predicted: 1900, the year that Lord Kelvin honored the Newtonian method in front of the assembled Royal Society in London. But in that same year, Albert Einstein graduated from the Zurich Polytechnic Academy. Max Planck published his first paper on the quantum, while Henri Poincaré started running into abstruse difficulties with Newtonian mechanics. This would lead to chaos theory more than half a century later.[14]

In the first decades of the twentieth century, several scientific disciplines started to show serious engagement with what we now call systems thinking. The emphasis on parts, as Boyle had pursued, Descartes had theorized and Newton had made lawful, began to show its limits in the explanatory power and development potential of various sciences at that time. Physics was, of course, one of them, and, particularly given its status as Descartes' "stem" in the system of sciences, quite a prominent one.

Relativity theory, developed almost entirely by Einstein himself, and quantum theory, which had a larger gaggle of physicists behind it, both served to change the view of physics forever – and with it our understanding of the universe. The theory of relativity captured why the world looks different to observers moving at different speeds and that there is no absolute frame of reference against which all speeds can be measured. For Einstein, though, fundamental laws of an objective nature underlie such relative appearances, a position that would later turn him into the "last classical physicist." Quantum theory, which grew out of classical experimental investigations into the make-up of atoms, revealed sensational and totally unexpected results. Far from yielding hard, basic, minimalist particles, the subatomic world was not populated by the solid objects predicted by Newton and classical physics:

> At the subatomic level the interrelations and interactions between the parts of the whole are more fundamental than the parts themselves. There is motion, but, ultimately, no moving objects; there is activity but there are no actors; there are no dancers, only the dance.[15]

Quantum physics showed that the Newtonian–Cartesian strategy of decomposition would, quite literally, lead to nothing. Decompose matter far enough, down to the subatomic level, and you are left not with basic particles, but with relationships, with mere probability patterns. Matter is not made out of smaller matter. It is made, really, out of "nothing" when you apply the terms of classical Newtonian physics. The parts cannot explain the whole, because the parts are basically not there. "At the subatomic level, the solid material objects of classical physics dissolve into wave-like patterns of probabilities. These patterns, furthermore, do not represent probabilities of things, but rather probabilities of relationships."[16]

Quantum physics showed that the universe could never be understood as a collection of isolated objects, and the entire notion of separate particles as the bearer of a fundamental explanation began to crumble. The most basic matter in the universe could not be explained as individual entities, but must rather be understood as complex webs of interrelationships, which were not mathematically predictable. A particle was not an independently existing, unanalysable entity. It was a set of relationships that reach outward to other things.[17] It began to dawn on scientists that laws and predictions could be expressed only as probabilities, not certainties. The outcomes of a complex system could not be predicted. The only thing that could be predicted was the approximate likelihood of something happening. Its probability, in other words.

One of my students recently argued that we really shouldn't be too pessimistic in our outlook for safety. Aviation, as a world-wide industry, is not drifting into failure. After all, he said, haven't we made enormous progress? Look at the accident statistics and the number of fatalities, he said. Of course, there are spikes and variations. But year over year, we seem to get better and better, not drifting into something worse.

Well, another said, this depends on how you define progress, and where you draw the boundary of the fatalities that the system may be thought to be responsible for (and, indeed, how you define such "responsibility"). We generate energy in aviation the very same way as we did in pre-history. We burn carbon-based material, just as we did when we were dwelling in caves. Doing the same thing for 10,000 years? We should not really count that as progress. What's more, how do we know that this is producing fewer fatalities? Doesn't that depend on where we look for system interconnections and where we start and stop counting dead bodies? The traditional way of assessing fatalities in aviation is to count the victims in the crashed airplane and any additional victims on the ground. That seems to make good sense, of course. But perhaps it is also a quite limited way of looking at the fatal consequences of flying. And, in a sense, a very parochial and self-indulgent one. We count only those in the world who were already fortunate (and affluent) enough to take direct part in the activity (which in itself is an increasingly sizable portion of humanity). But what about other people who are affected?

For example, we might want to look for connections between aviation activity, global warming, and increasing wars in Africa. The body count would go up dramatically, changing the notion of "progress" if only we are willing to include a bigger slice of humanity in the picture. Since 1990, emissions of carbon dioxide by aviation have risen 90 percent. According to some reports, aviation's share of CO_2 emissions could rise from about 2 percent globally today, to as much as 20 percent in 2050.[18] Of course, enormous economic opportunities are created by such global interconnections and their increase, in ways that are not very dependent on land-based infrastructure. So that could count as progress along a number of dimensions, for sure.

But there are other effects. The release of CO_2 into the atmosphere (each kilogram of aviation fuel that is burned basically adds two kilograms of CO_2 to the atmosphere) has been linked not only to a global increase in temperature, but to changes in regional climates as well.[19] The earth, of course, has always known temperature cycles and fluctuations in atmospheric gas composition. And only some of the CO_2 in the atmosphere comes from human activity. Since the dawn of the Industrial Revolution, however, this is an increasing (and, lately, accelerating) proportion. Today the coupling between anthropogenic CO_2 and climate change (or global warming) is seen as proven and lawful by many; others dismiss it as amusing and as junk science.

But changes in CO_2 concentrations (whatever their source) do have effects on regional climates. Using what they called a zonally symmetric synchronously coupled biosphere-atmosphere model, NASA and MIT scientists showed a connection between CO_2 concentrations and changes in the biosphere-atmosphere system in Africa. Changes in CO_2 concentration make certain regions wetter,[20] and, by extension, certain regions drier.[21] Other scientists have found a strong link between drier years and the prevalence of conflict in Africa. Long-term fluctuations of war frequency follow cycles of temperature change: there is more conflict during warmer years. Rain-fed agriculture accounts for more than 50 percent of gross domestic product, and up to 90 percent of employment, across much of the continent. A drier, warmer climate decreases agricultural output (a 10–30 percent drop per °C of warming), which in turn degrades economic performance, which in turn increases conflict incidence. This holds even after correcting for per capita income or democratization (increases in both tend to reduce war risk).

Warming, then, has a role in shaping conflict risk, with projected climate changes increasing armed conflict by 54 percent and bringing an additional 393,000 African deaths by 2030.[22] That would be like 20,000 deaths each year. If aviation's share of CO_2 production indeed rises from 2 percent to 20 percent over the next 40 years, its share in those deaths would drift from about 400 per year today to about 2,000 fatalities per year in 2030. In aviation, such figures would not count as progress; it could count as another slow, steady drift into failure. Also, it would outpace any beneficial effects of African economic growth and democratization (some of which, paradoxically, are also brought by aviation with its role in stimulating economic development and bringing people and ideas and goods and services together).

Of course, there are a lot of numbers in the example above. But the example is not about the numbers, it is about the possible relationships, and about where we believe we should draw the boundaries and call something a "system." Such boundaries are always exclusionary. Wherever we draw them, something will fall outside of it. Systems thinking is not saying that there should be no boundaries (because they are an essential part of defining what a system is), but that we can at least be more flexible, more imaginary, and perhaps more democratic in where we draw those boundaries.

And, as in all complex systems, other factors could play a plausible role too. All explanations for how system behavior is the result of relationships are tentative, and open for change when better arguments come in. It is not necessary, for example, that additional conflict deaths due to temperature increases are the result of regional warming. There is a known correlation between violent crime and higher temperatures, which could account for at least some of those additional deaths,[23] and even non-farm labor productivity can decline with higher temperatures, meaning that declines in agricultural output do not have to act as an intermediary variable between more heat and more deaths.[24] It is consistent with complexity thinking, then, when researchers say "We interpret our results as evidence of the strength of the temperature effect rather than as a documentation of the precise future contribution of economic progress or democratization to conflict risk. Similarly, we do not explicitly account for any adaptations that might occur within or outside agriculture that could lessen these countries' sensitivities to high temperatures, and thus our 2030 results should be viewed as projections rather than predictions."[25] Being modest about the reach of one's results, and explicitly open to their being wrong or open for revision, was once seen as a weakness in heavily naturalized science. But in complexity and systems thinking, it is a strength.

The example shows how physics was by no means left alone in this drift away from classical Newtonian science. In fact, systems thinking was in part pioneered by biologists, who began to look at living systems not as collections of components but as irreducible, integrated wholes. Social sciences and humanities have similarly put up ever stronger resistance over the past few decades against the wholesale naturalization of their research work. The only way to be "scientific" was once to be like physics (in fact, for centuries being more scientific meant having to be like physics). That meant that research had to be done in the form of carefully controlled experiments, in which the scientist had clear control over one or a few variables, tweaking them so that she or she or he could exactly understand what had caused what. Preferably, the results were expressed as, or converted into, numbers, so that the claim that something fundamental had been found about the world looked even more credible. This is the way of doing science that Newton and Descartes proposed. Ironically, with a shift of natural sciences into systems thinking and complexity, and social sciences and humanities doing the same, it seems that they once again seem to come closer to each other. Not because the social sciences and humanities are becoming "harder" and more quantifiable, but because natural sciences are becoming "softer" with an emphasis on unpredictability, irreducibility, non-linearity, time-irreversibility, adaptivity, self-organization, emergence — the sort of things that may always have been better suited to capture the social order.

It is interesting to see that complexity and systems thinking predates what we regard as the Scientific Revolution in the sixteenth and seventeenth centuries — a revolution that gave rise to the Newtonian–Cartesian hegemony. Leonardo (born Leonardo di Ser Piero, Vinci, Florence, 1452–1519) is regarded as science's first system thinker and complexity theorist. An artist, engineer, architect, relentless empiricist, experimentalist and inventor, Leonardo embodied the humanist fusion of art and science. He embraced a profound sense of the interrelatedness of things, interconnecting observations and ideas from different disciplines (for example, physiology and mechanical engineering; aerodynamics and music) so as to see problems in a completely new light and come up with nifty solutions.

Leonardo embraced his science as complexity – it had no boundaries since his goal was to combine, advance, investigate and understand processes in the natural world through an inter-disciplinarian view. He was a committed systems thinker, eschewing mechanistic explanations of phenomena and instead giving primacy to ecological ones – thinking up and out, rather than only down and in, and always seeing (and, in his art, depicting) linkages among living organisms (people, groups, organizations) so as to reveal new solutions. Leonardo was fascinated by the phenomenon of flight and produced many studies on the flight of birds (for example, his 1505 Codex). He designed a variety of aircraft – of which one, a hang glider, was recently demonstrated to actually work.

COMPLEX SYSTEMS THEORY

With his formulation of General Systems Theory, Von Bertalanffy helped establish a serious scientific foundation for the alternative to Newtonian–Cartesian thought in the early 1970s.[26] Today that alternative is known as complexity and systems theory. Recently, Cilliers summarized the characteristics of complex, as opposed to Newtonian, systems nicely.[27] I sum the points up as follows, and will explain them in more detail below:

○   Complex systems are open systems – open to influences from the environment in which they operate and influencing that environment in return. Such openness means that it is difficult to frame the boundaries around a system of interest.

○   In a complex system, each component is ignorant of the behavior of the system as a whole, and doesn't know the full effects of its actions either. Components respond locally to information presented by them there and then. Complexity arises from the huge, multiplied webs of relationships and interactions that result from these local actions.

○   Complexity is a feature of the system, not of components inside it. The knowledge of each component is limited and local, and there is no component that possesses enough capacity to represent the complexity of the entire system in that component itself. This is why the behavior of the system cannot be reduced to the behavior of the constituent components, but only characterized on the basis of the multitude of ever-changing relationships between them.

○   Complex systems operate under conditions far from equilibrium. Inputs need to be made the whole time by its components in order to keep it functioning. Without that constant flow of actions, of inputs, it cannot survive in a changing environment. The performance of complex systems is typically optimized at the edge of chaos, just before system behavior will become unrecognizably turbulent.

○   Complex systems have a history, a path-dependence. Their past is co-responsible for their present behavior, and descriptions of complexity have to take history into account.

○   Interactions in complex systems are non-linear. That means that there is an asymmetry between, for example, input and output, and that small events can produce large results. The existence of feedback loops means that complex systems can contain multipliers (where more of one means more of the other, in turn leading to more of one, and so forth) and butterfly effects.

Complex systems are not closed. They don't act in a vacuum like Newton's planetary system. Socio-technical systems or organizations are open systems. They are in constant interplay with their changing environment, buffeted and influenced by what goes on in there, and influencing it in turn. Whereas the systems studied by high reliability theory (see Chapter 4) were relatively closed (an aircraft carrier at sea, power grids before deregulation, air traffic control run by a single government entity, the Federal Aviation Administration), this is not true for many other safety critical systems. NASA, for example, has operated in a (geo-)political and societal force field for decades, suffused with budgets and operational priorities and expectations that never were her own but seeped in from a number of directions and players (congress, the defense department, other commercial space operators) and affected what were seen as rational trade-offs between acute production pressures and chronic safety concerns at the time. Of course the influence doesn't just go one way. NASA itself also affected what politicians saw as reasonable and doable, in part because of its own production, and its actions would have affected the actions of other commercial space operators too (for example, the choice by the French Ariane consortium not to use manned vehicles for the launching of satellites). This same openness goes for healthcare too: it often is a handmaiden of local or national politics and funding battles – hay gets made from promises to bring down surgical waiting lists, for instance. This can have effects on how funding is allocated, on where resources are added and where they are taken away.

Open systems mean that it can be quite difficult to define the border of a system. What belongs to the system, and what doesn't? This is known as the frame problem. It is very difficult to explicitly specify which conditions are not affected by an action. When you trace the ever-changing webs of relationships in the mining of Coltan for the production of cell phones, for example (see above), you will find that actions by local people reverberate across an almost infinite range of economies and ecologies, affecting not only wildlife in the eastern Congo, but everything from inflation of food prices in local villages surrounding Kahuzi-Biega, to the allure of discarded Russian cargo planes and those who remember how to fly them, to the volatility of commodity markets and demands of just-in-time production of electronics in Taiwan, to the prosperity of particular factions in the underworld of Belgium. And all of these reverberations produce further reverberations, in Russia, Belgium, Taiwan, affecting people and markets and all kinds of relationships there too.

One solution of complexity and systems theory is to determine the scope of the system by the purpose of the description of the system, not by the system itself. If the purpose is to describe the effects of Coltan mining on local gorilla populations, then you might draw the geographic system boundaries not far from Kahuzi-Biega. Then you can trace ecological contamination and physical habitat destruction, and you might restrict functional boundaries to bush-meat prices, hunting methods, and numbers of miner mouths to feed. Never mind Russia or Belgium or Taiwan. But such boundaries are a choice: a choice that is governed by what you want to find out. System theory itself provides no answer to the frame problem. Where you place the frame is up to you, and up to the question you wish to examine. The adage of forensic science to "follow the money" is the same commitment. On the one hand, it leaves the observer or investigator entirely open to where the trail may take her or him and how it branches out into multiple directions, organizations, countries. That is where the system of interest is open. On the other hand, the commitment frames the system of interest as that which can be expressed monetarily.

This highlights a very important aspect of the post-Newtonian perspective: the world we can observe isn't just there, completely and perfectly ready-formed and waiting for us to discover. We ourselves play a very active role in creating the world we observe, precisely because of our observations. If we don't have the language or the knowledge to see something, we won't ever see it. What our observations come up with is in large part up to us, the observers. In post-Newtonian science, it is very hard to separate the observer from the observed; to say where one ends and the other begins. In post-Newtonian science, reality does not contain the sort of hard, immutable facts that have a life entirely independent of a human observer. Human observation cannot be the neutral arbiter or producer of knowledge. After all, the observers themselves impose a particular language, interest, and they come with imaginations and a background that actively bring certain aspects of a problem into view, while leaving others obscured and unexamined.

In a complex system, each component is ignorant of the behavior of the system as a whole. This is a very important point.[28] If each component "knew" what effects its actions had on the entire rest of the system, then all of the system's complexity would have to be present in that component. It isn't. This is the whole point of complexity and systems theory. Single elements do not contain all the complexity of the system. If they did, then reductionism could work as an analytic strategy: we could explain the whole simply by looking at the part. But in complex systems, we can't, and analytic reduction doesn't work to enhance anybody's understanding of the system.

Complexity is the result of a rich interaction, of constantly evolving relationships between components and the information and other exchanges that they produce. Rich interaction takes a lot of components, which is indeed what complex systems consist of. Of course a beach has a lot of components (sand kernels) too, but that alone doesn't make it complex. The components have to interact. They have to give each other things like energy, information, goods. Those interactions, however, are limited, or local. Each component responds only to the limited information it is presented with, and presented with locally. The Coltan miner has no idea about the price of retired Russian cargo planes, and might not even recognize one if he saw it. What he mines is taken away through the bush on the backs of other men, perhaps with the use of animals or motorbikes, then onto trucks if they can make it that far into the jungle. He may not even know what he is mining, what the stuff is for, or why it is worth so much. The pilot flying the plane to a nearby airstrip is probably ignorant about gorillas, and may not know who really offloads his cargo at the receiving end. He makes the trip not because he works for a major cell phone manufacturer, but because he gets U.S. dollars in bundles of cash at the end of it.

Components in the system only act on and respond to information that is available to them locally. This locality principle again confirms the post-Newtonian view on knowledge and reality, as it makes access to some pre-existing, objective external reality not only impossible but also irrelevant. In the words of Heylighen, Cilliers and Gershenson:

> According to cybernetics, knowledge is intrinsically subjective; it is merely an imperfect tool used by an intelligent agent to help it achieve its personal goals. Such an agent not only does not need an objective reflection of reality, it can never achieve one. Indeed, the agent does not have access to any 'external reality:' it can merely sense its inputs, note its outputs (actions) and from the correlations between them induce certain rules or regularities that seem to hold within its environment. Different agents, experiencing different inputs and outputs, will in general induce different correlations, and therefore develop a different knowledge of the environment in which they live. There is no objective way to determine whose view is right and whose is wrong, since the agents effectively live in different environments ('Umwelts') – although they may find that some of the regularities they infer appear to be similar.[29]

That interactions between components in a complex system play out over fairly short ranges, and information is received primarily from immediate neighbors, does of course not mean that the ultimate influence of local agents is limited to those ranges. First, some agents physically travel through the system (like that pilots flying the cargo plan with Coltan), and will thus interact with multiple local settings along the way. Second, actions reverberate through further relationships, and thus affect what goes on in another part of the system. But because these reverberations pass through other components and their relationships, they can be modulated, suppressed, amplified.

THE BUTTERFLY EFFECT

One reason why this happens is because the interaction in complex systems is typically non-linear. It is non-linear not just in a mathematical sense (of equations whose terms are not of the first degree), and not in a spatial geometric sense (not arranged in a straight line), but rather in a physics sense, where non-linear means a lack of linearity or symmetry between two related qualities such as input and output. Non-linearity guarantees that small actions can eventually have large results. The amplification, or multiplication of effects in a complex, open system (a kingdom doing battle) was noted in a poem that dates back to at least the fourteenth century):

> For want of a nail, the shoe was lost.
> For want of a shoe the horse was lost.
> For want of a horse the rider was lost.
> For want of a rider the battle was lost.
> For want of a battle the kingdom was lost.
> And all for the want of a horseshoe nail.

As the interactions unfurled through the system, from nail to shoe to horse to rider to battle to kingdom, their effects spread and multiplied. Recognize how this contrasts with Newtonian science, which holds (according to the third law of motion) that the size of the effect is equal (and opposite) to the cause. Such symmetry is typically not the case in complex systems. This is not only related to nonlinearity, but to the nature of the feedback loops themselves, which in a complex system are typically not unidirectional. The effect of any action can feed back onto itself, sometimes directly, and sometimes through multiple intervening stages. When a favorite rider is out of action it affects surrounding riders, then surrounding units. The crumbling of morale through the ranks, then, could be one of those loops. The notion of multipliers has since then become a formal one in complexity theory.

Man-made CO_2 production could once be compensated by natural processes, but these may now no longer be able to keep up. As the oceans absorb CO_2, for example, they become more acidic. This, combined with increasing ocean temperatures, diminishes their ability to continue absorbing CO_2, and more CO_2 stays in the atmosphere. In 1960, a metric tonnne (1,000 kilograms) of CO2 emissions resulted in around 400 kilograms of CO_2 that remained in the atmosphere. In 2006, 450 kilograms of the same tonne remain in the atmosphere.[30] Hence a tonne of CO2 emissions today results in more heat-trapping capacity in the atmosphere than the same tonne emitted decades ago.

In a similar development, published in *Science* in 2010, it appears that methane coming out of the East Siberian Arctic Shelf (in addition to that coming from nearby Siberian wetlands) is comparable to the amount coming out of the entire world's oceans. The release of long-stored methane (a greenhouse gas more powerful than CO_2) is itself a result of global warming, which melts ice and permafrost and allows methane to bubble free from its ground storage (whether sub-sea or subterranean). In recent measurements, more than 80 percent of deep water in the East Siberian Arctic Shelf had methane levels between 8 and 250 times greater than background levels. That said, the water in the East Siberian Arctic Shelf isn't actually that deep, which means the methane coming up doesn't have enough time to oxidize, which means more of it escapes into the atmosphere. As it does, global warming could increase even more quickly, which in turn speeds up the release of more methane. This is a feedback loop, or a multiplier, where more means more. It is a feature of a complex system that is very hard to stop.[31]

The idea that small events can produce large results because of non-linearity is popularly known as the butterfly effect. The technical term, which was introduced in Chapters 1 and 2, is sensitive dependence on initial conditions. The phrase "butterfly effect" is related to the work by Edward Lorenz, a mathematician and meteorologist, and comes from the idea that a butterfly's wings might create tiny changes in the atmosphere that can ultimately balloon into a hurricane, or alter the path of a brewing hurricane, or delay, accelerate or even prevent the occurrence of a hurricane in a certain location. The flapping wing represents a small change in the initial condition of the system. This small change triggers a non-linear succession of events, leading to large-scale alterations of the state of the system. Had the butterfly not flapped its wings, the trajectory of the system might have been vastly different. Of course the butterfly does not "cause" the tornado in the sense of providing the energy for the tornado (indeed, that would be a Newtonian reading of "cause" with symmetry between cause and effect and a preservation of the total amount of energy through the causal chain). But it does "cause" it in the sense that the flap of its wings is an essential part of the initial conditions eventually resulting in a hurricane, and without that flap that particular hurricane would not have existed.

How small events can contribute to large results is a popular theme in the story of Anthony Eden, Foreign Secretary of the U.K. during the Suez crisis, and it is reminiscent of the loss of a kingdom (or empire) for want of a nail.[32] In 1952, a year after becoming Foreign Secretary in the new Churchill government, Anthony Eden suffered several attacks of upper abdominal pain, followed by more severe pain a year later. Sir Horace Evans, physician to the Queen, was called in and advised urgent surgery. On 12 April 1953 at the age of 55, surgery was performed in London by Mr John Basil Hume, surgeon at St Bartholomew's Hospital. While a very senior surgeon, Hume was not one with large amounts of recent experience. It is not entirely certain what happened during this operative procedure. But during cholecystectomy, Eden likely sustained an injury to the proximal common hepatic duct at its bifurcation and possibly, also, an injury of the right hepatic artery. In common terms, people would call it a botched gallbladder operation, which in itself is not uncommon.[33]

On 26 July 1956 Colonel Nasser seized the Suez Canal. This followed a long period of Egyptian opposition to what they regarded as a Canal Zone occupation, starting with King Farouk's demand for total and immediate withdrawal of British troops from the Suez Canal in 1950.

Ten weeks later, on 5 October 1956, Eden collapsed with a high temperature of 106°. This, more than three years after his bile duct repair, was the first of several major attacks of pain and fever.

Three weeks after his collapse on 27 October 1956, Israel invaded Egypt and a further 4 days later, Eden ordered the Anglo-French forces to occupy the Suez Canal Zone on the pretext of separating Egypt and Israel. In reality, the British government wanted the canal returned to international control to safeguard the oil supply of Western Europe, which passed through Suez. Eden also believed that Nasser might fall from power if he was forced to back down on his nationalization of the canal, and thus the Egyptian's brand of radical pan-Arab nationalism would be marginalized. But Eden's action caused an uproar in the British parliament, in the U.S.A and also at the United Nations. In response, all parties agreed to an early ceasefire on 6 November 1956, and 3 weeks after this occupation, on 23 November 1956, the Anglo-French troops began to withdraw from the Suez Canal Zone. Eden's action eventually had the geopolitical effect opposite to its intentions: he lost British control over Suez.

Prior to his collapse in July 1956, Eden was widely acknowledged by public servants working with him and by his many biographers, as a cool composed man, an expert in the use of diplomacy even under the most difficult circumstances. He believed in the rule of the law and in the supremacy and effectiveness of the United Nations. As Foreign Minister during Churchill's peacetime ministry, Eden had followed a global strategy that understood the post-war limitations on British power. Pursuing a military option by calling in the Anglo-French forces was, therefore, most uncharacteristic. At that time he was irritable, quick tempered, often tired, and most uncharacteristically, conspiratorial with France and possibly with Israel (although the latter has been denied). Uncharacteristically also, he upset the U.S.A., especially John Foster Dulles. Eden resigned 10 weeks later. He remained in poor health for the rest of his life.

ADAPTATION AND HISTORY

Complexity is a characteristic of the system, not of the parts in it. Recall that the Newtonian reflex – indeed, that of science in general – has been to see localized structures (for example, humans, or technical parts) as the prime causal agents. Without those, nothing gets organized, nothing happens. But complex systems do not rely on an external designer, and they do not rely for their continued functioning on some central internal coordinating agency or nerve center to specify how all the parts are going to work. Complex systems can keep on working in changing environments because the relationships among their parts can adjust themselves. These reconfigurations are a response to what goes on in the system, in the environment and in the multitude of interactions between them. Thus, complex systems are adaptive, and they can be resilient precisely because of their complexity. All the patterns of interaction between components make for a complex system. Webs of relationships wax and wane and adapt to what happens in the environment around it.

Drug cartels from South America have moved into West Africa, establishing themselves all over the region and opening up a new route for transporting cocaine from the plantations of Colombia, Peru and Bolivia to the consumers of Europe, particularly Britain and Spain. In recent years effective patrolling of traditional trans-Caribbean and transatlantic smuggling routes has forced the cartels to seek out new paths, which is where West Africa comes in: the shortest line of latitude westwards from the ports and airstrips of South America reaches land in Guinea-Bissau, which received the dubious honor of recently becoming West Africa's first "narco-state." Five years ago the amount of cocaine shipped to Europe via West Africa was negligible. Today 50 tonnes a year worth up to 2 billion dollars pass through the region. Interpol estimates as much as two thirds of the cocaine sold in Europe in 2009 alone will have reached the Continent via West Africa.

"We are seeing multi-ton shipments transiting West Africa. We have recorded arrests of Latin Americans all over West Africa," said Antonio Mazzitelli, at the regional office of the United Nations Office for Drugs and Crime in Dakar, Senegal. "They are using ships, speedboats, small and large airplanes, 4WDs. There is really no limit to the imagination of traffickers."

The cartels work with local criminal gangs and officials. Some consignments are for corrupt armies, customs and police forces. Plastic-wrapped cartons, each containing up to fifty 1-kg bars of cocaine, are divided up and distributed by gangs who use speedboats, trans-Sahara trucks, aircraft or human "mules," who board commercial flights to traffic the drugs to Europe. Spain is the usual entry point, with the northeastern Galicia region being the favorite place owing to the maze of inlets along its coastline. There is evidence that Irish gangsters, Balkan mobsters and Italy's Calabrian mafia are also involved.

The sheer value of drugs transported through West Africa – one of the world's poorest regions – dwarfs entire economies and corrupts security forces and politicians. Guinea is just one of more than a dozen countries in the region in varying states of disarray and poverty, many only recently emerging from years of bloody civil war fought over the control of "blood diamonds" and other resources. Cocaine, experts warn, is another resource that the gangs consider worth fighting for. "What we're seeing is the criminalization of the state as a result of drug trafficking," Corinne Dufka, a West Africa expert at Human Rights Watch, said. "The unlimited cash at the gangs' disposal risks toppling these desperately weak states."[34]

In the example above, individual traffickers seem very smart. They deflect routes away from patrols, and use a diversity of means of transportation. But it is unlikely that there is one single mastermind who has plotted it all out and who can keep up with plotting the right moves in response to a full knowledge of all the countermoves. Again, the mastermind would have to be as complex as the system that is being run. This is computationally impossible: the capacity for all that information (and its dynamics) exceeds what individual people or even computers can handle. Nor do they need to – that is the beauty of complex systems.

Rather, the traffickers in western Africa respond to local opportunities and constraints, just like the Latin Americans did by finding them as points of contact and throughput when flying or boating across the Caribbean became too risky. Local conditions drive local responses. If land-based traffic by 4-wheel drive vehicles gets difficult, speed boats will be used more to pace along the coast of West Africa. If policing becomes a little too effective on the northern plains of Guinea because of a surge of Interpol funding, then neighboring Guinea-Bissau becomes a hub instead, where local smugglers can make use of connections into the power elite and their military arms (as they did with the Red Berets in Guinea, once commandeered by the late President Conte in Conakry). That these patterns of adaptation play out in similar ways but at different scales (across the continents of Latin America and Africa, between the countries Guinea and Guinea-Bissau, across the different regions, towns, means of transportation and even fields or roads to use or avoid) is another property of complex systems. It is known as recursion (or fractals when speaking in geometric terms), where stochastically self-similar processes operate at different relative sizes or orders of magnitude. Such patterns, recognizable at different levels of resolution when studying complex systems, are one way of finding some order in how complexity works, some way of mapping what goes on there.[35]

Complex systems operate under conditions far from equilibrium. There has to be a constant flow of inputs to keep the system and its goals on track. Not making those inputs, those changes, means that the system will disintegrate, will stop functioning. In biological systems, equilibrium means death. For the smugglers of West Africa this goes too. If routes and methods are kept constant or stable, then they would be quickly shut down. It is precisely because they are only dynamically stable – able to function because of making inputs and responding to the environment the whole time, as if riding a bicycle – that they are able to survive in a changing world. They actively compensate perturbations that originate in the environment. Of course, by doing so, the environment will change too. The greater the variety of perturbations that a system has to respond to, the greater its diversity of responses has to become. It has to perform a greater variety of actions to stay in business.

If we look at the language that is often still used to describe activities like cocaine smuggling and their countermeasures, we can quickly see how it remains problematic and limiting. Calling it "organized crime," for example, is a convenient label, but it imports significant Newtonian assumptions. Organization (see also below under "organized complexity") suggests a form of Weberian bureaucracy: charts with boxes and connector lines between them, and somebody at the top of it. This does not reflect the highly dynamic, adaptive nature of how localized agents work, and how they collectively constitute a Coltan-mining and exporting "organization" of some kind – without even knowing that they do. And then, proclaiming that such an organization (a crime syndicate) has been dealt with by "chopping off its head" reduces the problem to what it isn't. Chopping off the head of a complex system doesn't work; it is even logically impossible. After all, its executive, its intelligence, its central nervous system, is distributed throughout the entire system. It is the system that is complex and smart and adaptive, not some omniscient governor that can be dealt with in isolation.

Of course, even in complex systems, not all components are equal. Some people are more involved in planning than others. And some people have a larger span of control or influence over other components than others. But that still doesn't mean that those people understand or control the whole system. Or that by taking them out of it, the system will collapse. This problem dogs the extrajudicial killings of terrorist leaders by remotely-operated drones too. Stopping terrorism by taking them out is really intractable, as others always step in to take their place, making up for their lack of experience with reinvigorated zeal. Announcing that one has interfered with "the machinery" of the criminal or terrorist organization and its activities also falsely suggests that componential solutions work. Taking one component out of a machine usually means it can no longer function. Taking one component out of a complex system does not mean it ceases to function. It adapts around the loss, making up for it, taking over functions and redistributing them. Importing Newtonian assumptions about the hunt for a culprit part by talking about the system in these terms, dramatically oversimplifies and mischaracterizes how these systems work; what makes them work, what makes them resilient.

Jim Nyce, an anthropologist colleague, and I have made this argument for the threat of improvised explosive devices (IED's) as well.[36] The U.S. Department of Defense (DoD) estimates that IED's are responsible for almost 50 percent of the casualties, mortal and injured, in Iraq to date and nearly 30 percent of the casualties in Afghanistan since the start of combat operations there.

Whenever we think about organizations and how to improve them, or try to understand how institutions in other cultures operate, we fall back on Weber, often without realizing it. This is not just because Weber's model informs much of our folk sociology; it is also because we take for granted that this model, based on what we take to be logic, rationality, and science, is the most effective and efficient organizational form. The taken-for-granted assumption becomes that if an adversary is effective in delivering IED's, in achieving its goals, then they have to have something like an efficient Weberian machine in operation. In other words, when we seek to understand IED financing, development, manufacturing, and distribution so as to stay ahead of the adversary's curve, we tend to look for some bureaucratic means of production, in terms of structure and hierarchy. Who is in charge where, of what, how are responsibilities managed, what are lines of supply, production? This tends to map directly on to questions of who to take out, which lines to disrupt.

But when something is not created bureaucratically, the means of organization and production, to say nothing of potential targets, tend to drop out of sight, and our means for determining appropriate action become less effective. All a Western analyst can do is register surprise that the original categories are not working, or that the adversary does not think like us, that he is less scrupulous, and just luckier.

The adversary's ability to adapt around our new technologies and better intelligence gathering does not hinge on a better bureaucracy on their part, even if some of their "coordination," may at first glance look very Weberian. What needs to be built into the equation is the possibility that these organizations work from and are informed by a set of principles fundamentally different from our own. This is often why these opponents, their intentions, and targets remain refractory, if not actually invisible, to us despite our best efforts. We do not have enough of the appropriate conceptual apparatus to see the groupings or entities that they do employ as organizations, let alone as threats.

Understanding how work gets done in the Middle East makes it clear that people there have refined the art of bricolage, of exploiting Western culture and its products, picking pieces from here and there, and then reassembling those in, for us, entirely incongruous, contradictory fashions. It is through this process of creative inspection, disassembly, and reassembly of Western technology and science that Iranian munitions become shaped charges, and garage door openers and cell phones become triggers for bombs. This also explains how our own media and analytic models, like network theory, have been turned against us. All this occurs through a set of kin relations, social institutions, and individual loyalties that defy any kind of Weberian description.

The West, however, consistent with the enlightenment and Newtonian science, tends to default to the individual. For this reason, terrorist acts are linked to something like demonic agency, as the ad hoc actions on the part of individuals that seemingly defy reason and morality. The IED threat, more than anything however, challenges the usefulness of applying such Newtonian–Cartesian concepts.

The example reveals another important point and that is that, in complex systems, history matters. Complex systems themselves have a history. They not only evolve over time, but their past is co-responsible for their present behavior. The IED threat faced by the various Western military forces in Iraq and Afghanistan is an example. U.S. funding and support for the Afghan Mujahideen (freedom fighters) during the 1979–1989 war against the Soviet Union there built the ideological and military basis for what they themselves have to battle against today. Known for their fierceness, and their stringent, extreme version of Islam, the Afghan Mujahideen attracted followers from a variety of countries who were interested in waging Jihad. Among those drawn to Afghanistan was a wealthy ambitious and pious Saudi named Osama bin Laden. He built what would later become al Qaeda, the militant Islamic fundamentalist group whose goal is to establish a pan-Islamic caliphate by expelling Westerners and non-Muslims from Islamic countries. Similarly, the accident of Alaska 261 cannot be understood without going back to the 1960s and how, from then on, the certification and rationalities surrounding the jackscrew and acme nut evolved. Any study of a complex system, then, has to be diachronic, that is, concerned with how it has developed over time. Its current behavior, after all, is path-dependent. Synchronic descriptions (that is, only a snapshot in time) will ignore a very important aspect of complexity.

All of these features of complexity mean that a complex system is never entirely describable. No set of equations can be drawn up that captures the entire functioning of the system. This has to do with the computational intractability and the instability of a complex system, and with its openness to a changing environment. The brain, social systems, "organized" crime, ecological systems – these can only be "understood" (though not in the Newtonian sense of fully described) as a system. This is precisely where reductionist methods fail. We have no direct access to the complexity. Our knowledge of complex systems is therefore of necessity limited.

COMPLEX VERSUS COMPLICATED

Complex is not the same as complicated. A complicated system can have a huge number of parts and interactions between parts, but it is, in principle, exhaustively describable. We can, again in principle, develop all the mathematics to capture all the possible states of the system. This is true of the Boeing 737 I fly, for example. Well almost – we will soon see why this is not completely so. It is a system with hundreds of thousands of parts that all interact in many ways. But these parts, and their interactions, can be described, modeled. This, of course, is not the same as saying that they *have* been described exhaustively. Or that any one human or computer program or simulator or flight manual can represent a model of everything that all parts of a 737 will do at all times in all environments and still enable people to understand it in its entirety. This probably exceeds the practically available computational capacity of most systems, including all pilots who fly the plane.

Complicated systems often (if not always) do rely on an external designer, or group or company of designers. The designers may not beforehand know how all their parts are going to work together (this is why there are lengthy processes of flight testing and certification), but in due time, with ample resources, in the limit, it is possible to draw up all the equations for how the entire system works, always. Reductionism, then, is a useful strategy to understand at least large parts of complicated systems. We can break them down and see how parts function or malfunction and in turn contribute to the functioning or malfunctioning of super-ordinate parts or systems.

Or can we? Two similar and hugely vexing accidents with 737s have happened in which it has been impossible to conclusively nail down the broken part (though a so-called Power Control Unit (or PCU) that helped govern the pressure applied on the rudder in the aircraft's tail, was always a suspect and indeed later replaced over initial manufacturer objections). These cases, where the aircraft seemed to suddenly flip on its back and then tumble straight down into the ground from low height, have been impossible even to meaningfully characterize at a system level. Was it the wind? Was it a rare, virtually unpredictable and very subtle interaction between autopilot rudder actuation, pilot action, and the wind? Was it the pilot pushing right rudder when left rudder was called for, or was it the pilot pushing left rudder but the airplane going right because of a PCU reversal problem?[37] The pulverized remains of the two aircraft yielded no answers, and even if the remains hadn't been pulverized, they would have likely yielded that answer in a Newtonian universe only. A universe which the scenarios of these accidents seemingly defied.

Perhaps complicated systems are not just complicated, but increasingly also complex. One wild card in this (growing in importance all the time) is software. Few parts or sub-systems of an airplane today (including very much the human pilots) do not somehow interact with software. This can mean that a system is no longer just complicated, but that it becomes complex. Of course, there is a designer behind software, though this is becoming recursive: software is increasingly written by other software. But that there was a designer at one point in time may mean very little once the system is fielded in an open environment.

Recall the example of TCAS or traffic collision avoidance systems from above, which are now a mandatory part of all large airplanes. With more than a million lines of code, they warn aurally and visually of other traffic and tell the pilot to maneuver up or down to avoid it. Telling the pilot to go sideways is outside the pale of the system. This would take considerably more computational capacity, millions more lines of code, and create considerably more uncertainty. With so much code already, and a literally infinite number of vector-to-vector situations, there really is no way to exhaustively model the interaction between all the parts of the code (for example, objects, lines) and all possible situations in which it may be triggered into action. Reductionism begins to lose traction here.

The apparently simple jackscrew-acme nut combination of Alaska 261 (Chapter 1) showed this nicely. Even for a very straightforward system, once the design was done, bets were off. Thread wear on acme nuts should, in principle, have been predictable with sufficiently accurate models of the amalgam of materials used to build the screw and the nut, of predictions of particle concentrations in the air (for example, salt, sand) in which airplanes were going to fly, and of the behavior of the lubricant over time, and over the likely temperature, air density and moisture ranges (and rates of change) along which all of this was going to rub together. But even if this was possible, or done, what could not be predicted were other parts of the environment in which it was going to operate. The system turned out to be quite open for influences way beyond the control or even imagination of those who designed it. Consider deregulation, developments in maintenance guidance and a philosophy towards more paper-based system-level oversight, evolutions of task distributions across different checks, operational ambiguities about on-airplane or off-airplane wear testing and the parts used to do it with, and how each individual operator of the airplane was going to interpret all of this and get it approved. None of this would have been easy to predict. It would have been impossible to model how it all might interact to produce legal lubrication intervals far beyond a 1960s engineer's worst nightmare.

Fielding an engineered system (like a PCU or jackscrew), that is, putting it to use in actual practical settings, opens a previously closed system. By opening it to the world, it will very likely start to display properties that are more akin to a complex, rather than complicated system. This, again, is the story of Alaska 261. Influences from regulatory relief, and the resulting scarcity and competition, and how this may have accelerated developments of system-level maintenance guidance and inspection recommendations start leaking into the system operating the jackscrew, and it responds in return. This once more shows the importance of diachronically understanding complexity: tracing the development of a system over time may reveal a lot about why it is behaving as it is synchronically, at any one point in time.

Complicated systems like a Boeing 737, or any other large aircraft (or, for that matter, almost any other complicated piece of technology) can also become complex when they are taken into cultural territories that lie at a far distance from the design and engineering assumptions that went into making it. Such implementation in unfamiliar or misunderstood social settings leads to unforeseen and uncontrollable explosions of relationships and interdependencies. Two seats in a cockpit and the artifacts (displays, switches, checklists) and procedures that connect them, are no longer animated by the manufacturer-envisioned, technical cooperation necessary to drive a piece of machinery alone. Putting people from cultures other than the manufacturer's in there breeds new relationships and interdependencies between people, procedures and technologies. It leads, in other words, to complexity.

Western assumptions about language competency, operational hierarchy and interpersonal communication and collaboration are woven into most technology produced in the Western world, even if implicitly. This has led to concerns, for example, that airliner cockpits and the way in which they are supposed to be operated collaboratively and relatively democratically today, are insufficiently sensitive to the cultural predispositions of some of the countries in which they end up being flown. Airliners that are designed and built in a culture that has no politeness differentiation in interpersonal address are flown in cultures that have such differentiation, like the two levels of many European languages, or like the six politeness levels of interpersonal address in some Asian languages. Western technology that is operated by multiple people also typically assumes a transmitter-orientation in interpersonal communication. That means that any lack of clarity (about content or urgency of the message) is mainly the responsibility of the sender, not the receiver. It is the sender who has to escalate, the sender who has to repeat, the sender who has to clarify. In some cultures, such transmitter initiative and insistence will be seen as violating and deeply impolite. And, independent of culture, the collaborative operation of machinery in time-critical situations assumes the use of unmitigated language – language that doesn't hedge, qualify, weaken or soften, but that delivers unvarnished and unambiguous messages. People generally have a really hard time doing that, because they are not the machine they are operating. They are people, with feelings and social expectations about relationships. Only constant training and indoctrination, as well as positive reinforcement, can help build confidence and acceptance for the use of unmitigated language.

There is no way that a technological environment can accommodate such nuance and diversity without consequences. Assumptions about purely technical cross-checking and challenging, baked into features of displays, procedures and checklists, fall flat when subjected to a culture that takes a different view of command, power, politeness and seniority.[38] The result is that original designs, when opened up to such influences, get adapted in ways that are hard to foresee by those who built it, by those who put the original parts together. The ability to describe, beforehand, the possible states that the system may get into, is challenged severely by the ways in which it will be put to use in different parts of the world.

> The importance of the diachronic study of complex systems comes out of a recent investigation into an accident with a Boeing 737 as well.[39] During approach, this aircraft suffered problems with its left radar altimeter. The radar altimeter is used in part to govern the autothrottles. If pilots have the intention to make an automatic landing, the radar altimeter will tell the autothrottles to start retarding 27 feet above the ground, and stay in retard. In this case, the right pilot (co-pilot) was flying the aircraft, and the right autopilot was switched on. The co-pilot might have thought that, as a result of this, the right radar altimeter was telling the autothrottles what to do, and that as such, he would suffer no consequences from the problems that the left radar altimeter was having. In all the books and training available to pilots on the 737, this is exactly the suggestion that is being made.

> But it is not true. The right radar altimeter never talks to the autothrottle. It is always the left radar altimeter that talks to the autothrottle, and tells it when to retard or do other things. This is independent of which pilot is flying, and independent of which autopilot is engaged. In the case of the accident, the left radar altimeter erroneously told the aircraft that it had already landed, even while it was still at 2,000 feet and a few miles short of the runway. The result was a loss of airspeed and eventual crash.

> The system design (originally certified in the 1960s) prioritized data integrity for the left pilot (commander), but sacrificed it for the first officer (second-in-command), consistent with societal and airline hierarchical arrangements at the time. In today's much flatter cockpits, however, it leaves a first officer with a seemingly autonomous system that is actually driven by "borrowed" data

and auto-throttle inputs from the left pilot, which can be corrupted without his knowledge. As airline training footprints shrank over the next decades, this information disappeared from pilot education and manuals, leaving an unknown automation booby trap long into the twenty-first century. A once complicated system became complex when exposed to the diverse, changing world in which it was to operate and survive.

COMPLEXITY AND DRIFT

The paradox is of course this. Complexity can guarantee resilience. Because they consist of complex webs of relationships, and because a lot of control is distributed rather than centralized, complex systems can adapt to a changing world. They can survive in the world thanks to this ability to adapt. So how can it be that complexity contributes to failure, to accidents? What is the relationship between complexity and drift? Complexity opens up a way for a particular kind of brittleness. Their openness means unpredictable behavior. Releasing a jackscrew into a world of competition and scarcity, and bets about maintenance intervals are off. Release a complicated engineered system into a world of cultural nuance, diversity and societal maturation, and original design assumptions get adapted, forgotten, muffled. But that is just the start. Complexity and systems theory gives us a language, and some metaphors, to characterize what may happen during the journey into failure, during the trajectory toward an accident.

The path-dependence of complex systems (or of a transformative journey from complicated to complex system) is a great starting point. Drift into failure can never be seen synchronically; systems have to be studied diachronically to have any hope of being able to discern where they might be heading and why. The non-linearity of relationships between components offers opportunities for dampening and modulating risky influences (for example, an increase in lubrication intervals might be accompanied by better endplay measurement devices and checking, which could increase the reliability of that part non-linearly even with less lubrication). But the non-linearity can also turn small events into large ones. The same small event, of missing one lubrication opportunity, would have had small consequences in the 1960s, and huge ones in the late 1990s. Making it increasingly difficult for smugglers to get their drugs across the Caribbean by numerous small steps that improved monitoring and interdiction led to a large event: a wholesale deflection of smuggling routes through West Africa.

In the mechanistic worldview, it is enough to understand the functioning or breaking of parts to explain the behavior of the system as a whole. In complexity and systems thinking, where nothing really functions in an unbroken or strictly linear fashion, it is not. Recall, from the second chapter, the outlines of drift into failure. Here is how they interact with what complexity theory has to say:

○ Resource scarcity and competition, which leads to a chronic need to balance cost pressures with safety. In a complex system, this means that the thousands smaller and larger decisions and trade-offs that get made throughout the system each day can generate a joint preference without central coordination, and without apparent local consequences: production and efficiency get served in people's local goal pursuits while safety gets sacrificed – but not visibly so;

○   Decrementalism, where constant organizational and operational adaptation around goal conflicts and uncertainty produces small, step-wise normalization where each next decrement is only a small deviation from the previously accepted norm, and continued operational success is relied upon as a guarantee of future safety;

○   Sensitive dependence on initial conditions. Because of the lack of a central designer or any part that knows the entire complex system, conditions can be changed in one of its corners for a very good reason and without any apparent implications: it's simply no big deal. This may, however, generate reverberations through the interconnected webs of relationships; it can get amplified or suppressed as it modulates through the system;

○   Unruly technology, which introduces and sustains uncertainties about how and when things may fail. Complexity can be a property of the technology-in-context. Even though parts or sub-systems can be modeled exhaustively in isolation (and therefore remain merely complicated), their operation with each other in a dynamic environment generates the unforeseeabilities and uncertainties of complexity;

○   Contribution of the entire protective structure (the organization itself, but also the regulator, legislation, and other forms of oversight) that is set up and maintained to ensure safety (at least in principle: some regulators would stress that all they do is ensure regulatory compliance). Protective structures themselves can consist of complex webs of players and interactions, and are exposed to an environment that influences it with societal expectations, resource constraints, and goal interactions. This affects how it condones, regulates and helps rationalize or even legalizes definitions of "acceptable" system performance.

The concern behind complexity and drifting into failure is how a large number of things and processes interact, and generate organizational trajectories when exposed to different influences. Resource scarcity and goal oppositions form one such pervasive influence. They express themselves in thousands of smaller and larger trade-offs, sacrifices, budgetary decisions – some very obvious, others hardly noticed. The ripple effects of such decisions and trade-offs are sometimes easy to foresee, but often opaque and resistant to anything resembling deterministic prediction. Decrementalism shows up in all kinds of subtle ways as people in the organization adapt, rationalize and normalize their views, assessments and decisions.

The contribution of the protective structure (for example, a safety regulator) to such adaptation and normalization, as well as exposure to its own resource constraints and goal interactions, is another influence on this. Such influences ebb and flow to different parts of the operational organization or even originate there, and are negotiated, dealt with, ignored or integrated. As Heisenberg put it, "The world thus appears as a complex tissue of events, in which connections of different kinds alternate or overlap or combine and thereby determine the texture of the whole."[40] Drift can be one property of all of these countless relationships and subtle interactions that emerges at the system level; as one aspect of Heisenberg's visible texture. Let's lift a few concepts out of that texture – emergence, phase shifts and the edge of chaos – to see how they can apply to drift into failure.

WHAT IS EMERGENCE?

As always in the tug between Newtonian and holistic systems of thought, the basic tension that animated this development was that between the parts and the whole. What was the relationship between parts and whole? Could the parts explain the whole, or did they fall short in accounting for the behavior of the whole? What could be considered as "parts" in the first place? In twentieth-century science, the holistic perspective, which rejects the idea that the whole can be understood as the sum of parts, has become known as systemic, and the way of reasoning it encouraged was called systems thinking.[41] Again, the basic reflex in systems thinking is to go up and out, not down and in. It is an understanding of relationships, not parts, that marks systems thinking.

The question of accidents in complex systems is most certainly a question about the relationship between parts and wholes. The Newtonian answer to the question of accident causation (and, by implication, the question of the relationship between parts and wholes) has always been simple: the parts fully account for the behavior of the whole. Hence our search for the broken part. Hence our satisfaction when we have found the "human error" by any other name or human, that can be held responsible for the accident. In complexity and systems thinking, the relationship between parts and wholes is – you guessed it – a bit more complex. The most common way to describe that relationship is by using the idea of emergence. A whole has emergent properties if what it produces cannot be explained through the properties of the parts that make up the whole. Here is an example:

What are the emergent properties of kitchen salt?

- It has a salty taste (no, really?)
- It is edible (well, in reasonable quantities)
- It forms crystals

So what are the components that make up kitchen salt? You might say: salt crystals. Well, no. That's the whole point. The crystals are an emergent property, as are the taste and their edibility. The "components" that kitchen salt is made up of are Na and Cl, or Sodium (Natrium) and Chlorine. Sodium is a poisonous gas. Natrium is a violently reactive soft metal (Yes, you eat this. A lot). So kitchen salt displays properties that are completely different from the properties of its (chemical) component parts. And what's more, it doesn't display the properties of its component parts at all. You could argue that it only gets dangerous at quantities like those consumed by Morgan Spurlock in his 2004 documentary "Supersize Me," not because it's a poisonous gas or a violently reactive metal but because it does things with blood pressure and kidney function and such. You remember Morgan puking out of the car window after eating yet another Mackey D's breakfast, the camera gleefully tracking the yellow glop as it spluttered onto the pavement? Okay, that's not what I'm talking about.

We used to say that the whole is more than the sum of its parts. Today, we would say that the whole has emergent properties. But it means the same thing. Emergence comes from the Latin meaning of "bringing to light." The nineteenth century English science philosopher George Henry Lewes was probably the first who made a distinction between resultant and emergent phenomena.[42] Resultant phenomena can be predicted on the basis of the constituent parts. Emergent phenomena cannot. The heat of apple sauce is a resultant phenomenon: the faster the constituent molecules are moving around, the hotter the apple sauce. The taste of an apple, however, is emergent. Taste does not reside in each individual cell, it cannot be predicted on the basis of the cells that make up the apple. Wetness does not reside in individual H_2O molecules. They are not wet, they don't flow. Wetness is an emergent property, something that can only be observed and experienced at system level. Wetness cannot be reduced to (or found in) the molecular components that make up water.

Theorizing around emergence took off in earnest in the late sixties with the study of slime mold in New York City.[43] Why slime mold? Because as a collective, or a whole, it does some amazing stuff that the component parts couldn't dream of pulling off. Slime mold is like ordinary fungi, or mildew. It's like the stuff on the inside your basement walls. Mushrooms are fungi too, a kind of mold in another form. Not long ago, a Japanese scientist succeeded in getting slime mold to find its way through a maze, and have it find food. One clump of slime mold even stretched itself thin to gain access to two sources of food in different places in the maze simultaneously. In a later experiment, it was possible to have the slime mold stretch itself out in a mimic of the Tokyo subway map.

Sending organisms through mazes in the pursuit of food is something that scientists used to reserve for mice and rats and other rodents. What puts rodents apart from slime mold is the same thing that separates a rule-based system from a complex system: they have a central nervous system – a location for their supposed intelligence, a mechanism for them to learn about and map their world, to build some kind of memory, and to then act in accordance with it (for which rodents have four paws, to walk them through the maze). Slime mold doesn't have a central nervous system. It also doesn't have paws. In fact, slime mold doesn't have much of anything, other than basic biological material that can reproduce and that needs food to survive. Slime mold has no central organizing entity, like a brain, that connects eyes on the one side with paws on the other. Nothing or nobody is in charge, nothing or nobody directs eyes where to look, paws where to walk. There is no specialization of parts that take on the various tasks. Every cell or protoplasm of slime mold is basically the same. One big egalitarian democracy of jelly, or slime.

In other words, slime mold is made up of very simple components, but it does very complex stuff, in interaction with its changing environment. Its emergent behavior would seem intelligent. The slime mold as a collective, amazingly enough, pursues goals. Like getting food. And collectively, it adjusts its behavior on the basis of what it finds in the environment. It does this really well. Slime mold actually walks. Well, in a sense. It moves across the forest floor, for example, in pursuit of food. And it gets even more uncanny: if there's a lot of food, it will actually break up into a flock, so as to capitalize on the various food sources. If there's not a lot of food, slime mold will pull together into one larger clump and stick together to ride out the lean times.

ORGANIZED COMPLEXITY

Such emergent behavior relies on a kind of distributed, bottom-up intelligence, not on a unified, top-down intelligence. Rather than running one single, smart program (as the rat in the maze supposedly would), there is a whole swarm of simple mini-programs, each running inside the cellular or a-cellular components that make up the mold. These programs really are sets of very simple rules. Anthills get built like this. In fact, much of ant colony life is organized around a few sets of simple rules. To find the shortest route to food, for example, (1) wander around randomly until finding food, (2) grab it and follow your pheromone trail back to the nest. Other ants will follow the same trail out because it is strong and fresh. If they don't find food at the end of it, or the trail begins to evaporate, the ants will go back to the first rule. Anthills, which are very sophisticated emergent structures, get built in the same way, though the rules are slightly different. Even the ant garbage dump and the location of their cemetery (maximal distance from the anthill) emerge from the millions of interactive applications of a few simple rules. Even the biblical book Proverbs from around 950 BCE, containing maxims attributed mainly to Solomon, refers to the ants. They have no king, no central authority, Proverbs says. Yet they are industrious and build great things.

What emerges has been called "organized complexity." It is complex because there are a large number of components, and, as a result, a dense throng of mini-programs running and interacting, cross-influencing each other. Such complex interaction and constant cross-adaptation is hard to model and understand using traditional deterministic, or Newtonian logic, because so much of it is non-linear and non-deterministic. But what it produces is not disorganized. Rather, it is organized: it creates higher-order behavior (like wandering in pursuit of food, or a huge ant stack that heats and cools and stores and incubates) as an amazing emergent product of the complex interactions between a multitude of simpler entities. Of course, the word "organized" is perhaps a bit problematic in this context, as it suggests some kind of entity that does the organizing or controls it. Which is not the case. The organization is purely emergent, purely bottom-up, the result of local interactions and the cumulative effect of simple rules.

Behavior of the brain has also been seen as emergent. In fact, it is both ironic and inspiring that we no longer think of the mammalian brain, the central nervous system, as a centralized smart program either. Rather, brains are made up of a huge mass of much simpler components with simpler programs (neurons or nerve cells) that, in their action-reaction or even ON–OFF behavior cannot begin to show the sort of complex phenomena that a collective brain is capable of exhibiting. Indeed, one example of emergence that is often used is consciousness. The cells that make up the brain are not conscious (at least not as far as we know. And, for that matter, how could we ever know?). But their collective interaction produces something that we experience as consciousness. Of course, much of human factors, thanks to its reliance on information processing psychology, still treats the brain as a central executive function: as an entity that produces and directs intelligent behavior. But the components that make up the brain are hardly intelligent. Its intelligence cannot be reduced to its constituent components. Intelligence itself is an emergent property.

ACCIDENTS AS EMERGENT PROPERTIES

Since ideas about systemic accident models were first published and popularized, system safety has been characterized as an emergent property, something that cannot be predicted on the basis of the components that make up the system.[44] Accidents have also been characterized as emergent properties of complex systems.[45] They cannot be predicted on the basis of the constituent parts; rather, they are one emergent result of the constituent components doing their normal work.

But, you may object, isn't there a relationship between some components (people, or technical parts) not doing their work, and having an accident? Indeed, if you believe this, it would affirm the idea that accidents are resultant phenomena, not emergent. After all, the fact that an accident happens can be traced back, or reduced, to a component not doing its job. The accidents-as-resultant-phenomena idea is alive and seems intuitive, consistent with common sense. In Alaska 261, for example, the component that didn't do its job was the jackscrew in the tail of the MD-80 airplane. It broke and that's why there was an accident. In Tenerife, one of the components that didn't do his job was the co-pilot, who didn't speak up against a stubborn captain.[46] That human component didn't work, and that's why there was an accident. Such characterizations are quite popular and, in many circles, still hard to call into question.

Those who are involved with occupational health and safety issues may want to believe it too. For example, isn't there a relationship between the number of occupational accidents (people burning themselves, falling off stairs, not securing loads, and so on) and having an organizational accident? Isn't it true that having a lot of occupational accidents points to something like a weak safety culture, which ultimately could help produce larger system accidents as well? Not necessarily, because it depends on how you describe the occupational accidents. If accidents are emergent properties, then the accident-proneness of the organization cannot be reduced to the accident-proneness of the people who make up the organization (again, if that is the model you want to use for explaining workplace accidents). In other words, you don't need a large number of accident-prone people in order to suffer an organizational accident. The accident-proneness of individual employees fails to predict or explain system-level accidents. You can suffer an organizational accident in an organization where people themselves have no little accidents or incidents, in which everything looks normal, and everybody is abiding by their rules.

The Alaska Airlines 261 accident, the topic of Chapter 2, was, as far as is known, not preceded by or correlated with a noticeable number of relevant occupational accidents. In fact, what is remarkable about this accident is that everybody was pretty much following the rules. Their local rules. The airline and its maintenance arm was abiding by recommendations from the manufacturer and the rules set by the regulator, the regulator was approving successive maintenance intervals on the basis of the evidence that was presented and deemed appropriate and sufficient at the time. The people doing the maintenance work followed the rules and procedures (however underspecified they might have been, but that was normal too) given to them for the execution of the lubrication task. Whether more or fewer maintenance workers would have suffered occupational incidents doesn't matter. It has no predictive value, it can carry no explanatory load relative to the eventual accident that emerged precisely because everybody was following their local rules. Just like the ants building their hill.

In Alaska 261, whether fewer or more technicians' hands got stuck and injured in the lubrication of the jack screw may bear no relationship to the eventual organizational accident. That is emergence. The behavior of the whole cannot be explained by, and is not mirrored in, the behavior of its constituent components. Instead, the behavior of the whole is the result – the emergent, cumulative result – of all these local components following their local rules and interact with each other in innumerous ways as they do so. Of course, unlike ants and slime mold, people across various organizations are not all the same. They are, in fact, quite different. But, like ants, all of them still respond to, and help shape, their local environment and follow or help shape the rules (written or otherwise, formal or informal) that apply there. So the properties of emergence would seem to hold. Behavior that is locally rational, that responds to local conditions and makes sense given the various rules that govern it locally, can add up to profoundly irrational behavior at the system level. Like an accident that takes scores, or hundreds, of lives – in a system where everybody locally works precisely to prevent that from happening.

A recent example from a major energy company operating in Louisiana near the Gulf of Mexico illustrates the problematic connection between the safety of individual components and system safety.[47] An employee of this company injured himself when his swivel chair collapsed. The injury was severe enough for him to have to take a day off work. People from Occupational Health and Safety (OSHA) came in, inspected the situation, and issued a citation to the company for "not properly instructing employees how to sit in their chairs."

What was the company to do? Under pressure to show managerial action in response to the OSHA finding and citation, their health and safety department sent out a company-wide PowerPoint presentation, which demonstrated how to properly and safely sit in a swivel chair. In summary, it told employees to inspect their chair at least once every month, and to remove defective chairs from the workplace. Also, employees were to use the chair only for the purpose for which it was designed, and never stand in a chair to retrieve an object out of reach. The company president himself got personally involved with enforcing the OSHA sitting recommendations, citing employees who violated the new policy. His rationale was that "Permitting the smallest exceptions to our health, safety and environmental program is unacceptable and results in catastrophes such as the BP disaster."

Chairs tipping over because people don't sit in them in ways that they were designed for may have a relationship with a disaster the size of Deepwater Horizon, but it may not. Complexity doesn't allow us to draw such straight lines from the behavior of individual components to large events at system level. In fact, complexity understandings tipping to mean something different.

PHASE SHIFTS AND TIPPING POINTS

The emergent behavior of a complex system can be very different, from one situation or moment to the next, than the collection of its parts would suggest. One dynamic that is responsible for this is called the tipping point, the phase shift, or phase transition. A phase shift means that a bit more (or less) of the same leads to something *very* different. Of course, if the relationship between parts and system were as straightforward as Newton proposed, then this could never happen. A bit more behavior by the parts would lead to the same little bit more behavior by the system as a whole. But in complexity, that is not the case. A tiny little bit more (or a tiny little bit less) of the same in terms of the parts, can lead to something *completely* different for the whole system. It can lead to something qualitatively different.

The original idea for phase transitions comes from physics (or, more specifically: thermodynamics). It describes the shift of a thermodynamic system from one phase to another. As solid material is heated, it will transition to a liquid at some temperature, shifting phases, and changing system level properties. Heat it more, and it will change to a gas. Heat it eve more, and it may, in rare cases, become plasma. The behavior of the parts (molecules moving among each other) is not at all that different on either side of a phase transition point, yet the system properties undergo an abrupt change. For example, the volume taken up by steam is vastly different from the volume that boiling water needs (yet the difference in temperature between these two only needs to be infinitesimal).

The idea gained currency in sociology in the 1960s. The term tipping point, or angle of repose, was introduced to describe how a previously rare social phenomenon could become rapidly and dramatically more common. Morton Grodzins was studying racially diverse neighborhoods in the U.S.A. in the 1960s, when he discovered that most of the white families remained in the neighborhood as long as the comparative number of black families remained small. Very small. At a certain point, however, when "one too many" black families arrived, the remaining white families would move out en masse, in a process that became known as "white flight." The phrase tipping point of course was borrowed from physics itself, analogous to the adding of a small weight to a balanced object that could cause it to suddenly and completely (and irreversibly) topple over.

On September 17, 2007, in Nisour Square, Baghdad, 17 people were killed and 24 injured when security teams from the private firm Blackwater U.S.A. unleashed a barrage of machine-gun fire. A bomb had gone off nearby just before. The first victim was Ahmed Haithem Ahmed, who was driving his mother, Mohassin, to pick up his father from the hospital where he worked as a pathologist. A Blackwater bullet tore through his head, but the car kept rolling toward the Blackwater convoy, and not much later 17 people were dead in a hail of gun fire, including Iraqis trying to escape to safety. No shot that could have provoked the Blackwater response had been heard.[48]

The event blew the extent of the privatization of warfare in Iraq into the open, with some in the U.S. congress arguing that if war was a profit motive, then peace could be hard to achieve. Blackwater lost its lucrative State Department contract to provide diplomatic security for the U.S. embassy in Baghdad after the incident.

Yet it would take two more years for the subtle drift towards the deadly incident to become apparent.

Blackwater U.S.A., a private security firm, was originally contracted to provide security for State Department and CIA employees after the September 11 attacks in 2001. In the spring of 2002, Erik Prince, the founder of Blackwater, offered to help the CIA guard its makeshift Afghan station in the Ariana Hotel in Kabul. Not long after signing that contract, dozens of Blackwater personnel, many of them former Navy Seals and Army Delta Force ex-soldiers, were sent out into the surrounding streets to provide perimeter security for the CIA station. From there, Blackwater operatives began accompanying CIA case officers on missions beyond the perimeter.

A similar progression happened in Iraq. Blackwater was first hired to provide static security for the CIA Baghdad station. Also, Blackwater employees were hired to provide personal security for CIA officers whenever they traveled in either Iraq or Afghanistan. This meant that Blackwater personnel began to accompany CIA officers even on offensive operations, sometimes launched in conjunction with Delta Force or Navy Seals teams (that is, former colleagues). It will never be possible to find out who fired the first offensive, not security-defensive, shot from a Blackwater gun in these operations, but lines soon began blurring.

Blackwater employees started to play central roles in so-called "snatch-and-grab" operations, intended to capture or kill militants in Iraq and Afghanistan, particularly during the 2004–2006 height of the Iraqi insurgency. Blackwater exercised a strong influence on such clandestine CIA operations, under the banner of being able to decide what the safest ways were to conduct such missions. They filled all roles, from "drivers to gunslingers."[49]

The House Oversight and Government Reform committee found that Blackwater had been involved in at least 195 shootings over the previous two years, many of which involved cover-ups of fatal shootings by its staff.

The incident on the 16th of September, 2007, was perhaps not a large departure from where things had been drifting. The Blackwater convoy in question was in the square to control traffic for a second convoy that was approaching from the south. The second convoy was bringing diplomats who had been evacuated from a meeting after a bomb went off near the compound where the meeting was taking place. That convoy had not arrived at the square by the time the shooting started.

As the gunfire continued, at least one of the Blackwater guards began screaming, "No! No! No!" and gestured to his colleagues to stop shooting, according to an Iraqi lawyer who was stuck in traffic and was shot in the back as he tried to flee.[50]

In the blur between CIA, military and contractor roles that grew during the worst part of the 2004–2006 Iraqi insurgency, not much would have distinguished one "snatch-and-grab" raid from the next. Blackwater guards would have fired shots in defense of the CIA and military, consistent with their assignment. Until a shot was fired that was not in defense, or not entirely in defense, or not at all in defense, but rather a contribution to the offensive raid. It is not that hard to imagine. You know the guys by name, you remember them from the time in the Seals or in Delta, you go on raids side-by-side for weeks, you see them lift a gun and fire in the same direction that everybody else is shooting in, you even get a shout perhaps, or a taunt, or a question. You shoot too. What's one offensive shot between friends? The next raid, you may even be expected to help out that way. One more shot, a bit more of the same, and the system began to display vastly different properties. Blackwater started to go in ahead of the others, started to help plan operations and take the lead, started to play an offensive role in the missions. Did the killing, even when unprovoked. A first offensive shot may have been a tipping point.

Despite original design requirements that the External Tank not shed debris, and the corresponding design requirement that the Shuttle not receive debris hits exceeding a trivial amount of force, debris impacted the Shuttle on each flight. Debris strikes were normal, in other words. Just like a lot of other technical problems – NASA engineers were, and always had been, working in an environment where technical problems proliferated. Flying with flaws was the norm. Over the course of 113 missions, foam-shedding and other debris impacts came to be regarded less as a hazard to the vehicle and crew. With each successful landing, it appears that NASA engineers and managers increasingly regarded the foam-shedding as inevitable, and as either unlikely to jeopardize safety or simply an acceptable risk.

The distinction between foam loss and debris events also appears to have become blurred. NASA and contractor personnel came to view foam strikes not as a safety of flight issue, but rather a maintenance, or "turnaround" issue. In Flight Readiness Review documentation, Mission Management Team minutes, In-Flight Anomaly disposition reports, and elsewhere, what was originally considered a serious threat to the *Orbiter* came to be treated as "in-family," a reportable problem that was within the known experience base, was believed to be understood, and was not regarded as a safety-of-flight issue. The reason why this problem was in the known experience base was that its result, heat tile damage, had occurred on previous occasions (in fact, was very normal) and occurred because of a variety of reasons. Here was just one more.

The foam-loss issue was considered insignificant enough that Flight Readiness Review documents included no discussion about it. There was no paper trail of concerns about foam debris tile damage that preceded the accident. This even fit the rules. According to Program rules, this discussion was not a requirement because the STS-112 incident was only identified as an "action," not an In-Flight Anomaly. Official definitions were assigned to each in-flight anomaly and ratified by the Flight Readiness Reviews. It limited the actions taken and the resources available for dealing with these problems.[51]

In the evaluation of damage caused by debris falling off the external tank prior to the 2003 Space Shuttle *Columbia* flight, you can see a similar phase shift. Under pressure to accommodate tight launch schedules and budget cuts (in part because of a diversion of funds to the international space station), it became more sensible to see certain problems as maintenance issues rather than flight safety risks. What was known as "debris events" now became "foam loss," a more innocuous label. Maintenance issues like foam loss could be cleared through a nominally simpler bureaucratic process, which allowed quicker turnarounds. In the enormous mass of assessments to be made between flights, foam debris strikes became one more issue. Gradually converting this issue from safety to maintenance was not different from a lot of other risk assessments and decisions that NASA had to do as one Shuttle landed and the next was prepared for flight. It was quite normal. It may, however, have represented the kind of phase shift, or phase transition – one more decision, just like tens of thousands of other decisions, that produced fundamentally different system behavior in the end.

With the benefit of hindsight, of course, it is easy to point to the flaws in these logics and priorities, for example those that converted a flight safety problem into a maintenance problem. But what we really need to understand is how these conversions of language made sense to decision-makers at the time. After all, their objective cannot have been to burn up a Space Shuttle on re-entry. And the important question to ask ourselves is how organizations can be made aware early on that such shifts in language can have far-reaching consequences, even if those are hard to foresee. In complex systems, after all, it is very hard to foresee or predict the consequences of presumed causes. So it is not the consequences that we should be afraid of (we might not even foresee them or believe them if we could). Rather, we should be weary of renaming things that negotiate their perceived risk down from what it was before.

OPTIMIZED AT THE EDGE OF CHAOS

A common notion is that the functioning of complex systems is optimized at the edge of chaos. Originally a mathematical concept, the edge of chaos in socio-technical settings denotes a region of maximal complexity where systems are perched between order and disorder; between optimally organized complexity and chaos. Maximal functionality can be extracted from the system at this edge. This is where the system is tweaked to achieve its extreme diversity and variety, where complexity reaches its optimum. Here, the system can maximally, exhaustively and swiftly adapt and respond to changes in the environment (4WDs to speed boats to airplanes, day to night, Caribbean to Guinea). Indeed, this optimum, or maximum is determined very much in the relationship between complex system and environment.

In a sense, the edge of chaos is where the ecological arms race plays out in its most pure form – where competitors, or predator and prey are constantly trying to stay one step ahead of each other. The use of aerial surveillance will make trafficking over open ground less attractive, perhaps deflecting smugglers' routes into forests. The use of infrared or other dark-penetrating optics makes trafficking at night less protected, thus putting a greater premium on quiet daytime travel, when the sun is hot and high and policing might be less effective. The various actions and counteractions constantly affect each other to undermine what was previously a good strategy while at the same time forcing the creation of new strategies. Complex systems tend to settle at the edge where their responses are just good enough to stay ahead of the others, but where there is not such a huge cost to generating those responses that they will run out of business because of it. That is the edge of chaos.

There is a fundamental trade-off that interdependent agents in a complex system must make, a trade-off that lies behind much of the adaptation that the system can display, and also why such adaptation sometimes becomes less successful. This is the trade-off between exploitation versus exploration. Exploration is a necessary activity for survival inside a complex, adaptive, living system. It means searching for a solution that is optimal relative to the landscape of opportunities and constraints you now know. Flying cocaine via West Africa may be a great opportunity that can be the result of exploration if the Caribbean has gone solid with narcotics countermeasures, policing, interdiction and arrests. Exploration is that which generates new smuggling routes (for example, through Africa). But making those routes work takes investment, of time and money. Local politicians may need to be bought, local strongmen need to be found and patronized, the local geography and ecology of inlets, ports, airports, roads, forests, hiding places, mules and so forth, need to be mapped. So if there are still plenty of holes to wiggle cocaine through in the Caribbean, then West Africa may not be optimal. The Caribbean can still be exploited – until further.

Exploitation means taking advantage of what you already know, it means reaping the benefits of past searches.[52] If you stop exploring, however, and just keep exploiting, you might miss out on something much better, and your system may become brittle, or less adaptive. While you are exploiting Caribbean routes, competitors may have already set up shop in Guinea. By the time the Caribbean holes close, you are left with no alternatives other than declaring war on rival smugglers. At the edge of chaos, then, the system has reached an optimum point – not just in either exploration or exploitation, but in their complement. It gets enough return from exploiting that which was explored earlier, yet retains adaptive capacity by continued exploration of better options. In a complex system, however, it is difficult to say in advance what the returns of exploration are going to be. A new route may get discovered. But a key smuggler on a scouting mission may get caught, and give up his or her collaborators. Exploration can thus produce big events. This is because the optimum balance between exploration and exploitation puts a system near the edge of chaos.

The edge of chaos is what is called a critical state, or, in an explicit reference to thermodynamics, a dissipative state. While poised in a critical or dissipative state, a system is not in equilibrium and can be easily perturbed. In this state, the many components that make up the complex system never quite lock into place, yet hardly ever dissolve into an entire chaos. For example, by purchasing a faster Coast Guard aircraft to go after jet-bound smugglers across the Caribbean, there will be a response. The smugglers' jet may fly at night, or change routings more often. Perturb the system again, and there might be a completely different response, and one at a different scale – such as the opening of routes through West Africa. This is one very characteristic property at the edge of chaos: Small responses are common, big responses are rare. But big responses are possible, and what they are can be almost impossible to foresee.

Arriving at the edge of chaos is a logical endpoint for drift. At the edge of chaos, systems have tuned themselves to the point of maximum capability. Mixing the base chemicals in a cup and smearing them across scratches and cracks and gouges in the foam covering of the Space Shuttle's external fuel tank was one such response to production pressures. Those actions got the tanks returned to service quickly by making evidence of any maintenance and possible flight safety problems go away. Production was served, and production had been called for (incentivized in various ways, even). The system had tweaked itself to maximum capacity in a constant, dancing interaction with the political, medial and technical environment in which it operated. It was in a critical state, which suddenly allowed a big response to the small, little actions that maintenance workers had applied hundreds of times. A large event happened on 16 January 2003: the separation of a chunk of iced-up foam so large that it managed to penetrate the heat shield on the wing's leading edge, which would later lead to the loss of *Columbia* while on its re-entry. Drift into failure, in these terms, is about optimizing the system until it is perched on that edge of chaos. There, in that critical state, big, devastating responses to small perturbations become possible. Large events are within the space of possibilities. Drift doesn't necessarily lead to failure. At least not until it does.

REFERENCES

1   Laszlo, E. (1996). *The systems view of the world: A holistic vision for our time*. Cresskill, NJ: Hampton Press.

2   Perrow, C. (1984). *Normal accidents: Living with high-risk technologies*. New York: Basic Books.

3   Perrow, C. (1984). Ibid.

4   Perrow, C. (1984). Ibid.

5   Seaways. (2010, July). London: The Nautical Institute, pp. 25–7.

6   See, for example, Billings, C.E. (1996). *Aviation automation: The search for a human-centered approach*. Hillsdale, NJ: Lawrence Erlbaum Associates.

7   Leveson, N.G. (2006). *System safety engineering: Back to the future*. Cambridge, MA: Aeronautics and Astronautics, Massachusetts Institute of Technology.

8   Mackall, D.A. (1988). *Development and Flight Test Experiences with a Flight-Critical Digital Control System* (NASA Technical Paper 2857). Lancaster, CA: National Aeronautics and Space Administration Dryden Flight Research Facility.

9   Leveson, N.G. (2006). Ibid.

10   Leveson, N.G. and Turner, C.S. (1993). An Investigation of the Therac-25 Accidents. *IEEE Computer*, 26(7), 18–41.

11   Rose, B.W. (1994). *Fatal dose: Radiation deaths linked to AECL computer errors*. Montreal, QC: Canadian Coalition for Nuclear Responsibility.

12   Leveson, N.G. (2006). Ibid.

13   United Nations Security Council (1). *Security Council, 4317th and 4318th meeting, condemns illegal exploitation of Democratic Republic of Congo's natural resources* (UN Press Release SC/7057). New York: United Nations.

14 Peat, F.D. (2002). *From certainty to uncertainty. The story of science and ideas in the twentieth century.* Washington, DC: Joseph Henry Press.

15 Capra, F. (1982). *The turning point.* New York: Simon & Schuster.

16 Capra, F. (1982). Ibid., p. 69.

17 Stapp, H.P. (1971). S-matrix interpretation of quantum theory. *Physical Review D*, 3.

18 Griffiths, P. (2009, September 8). *Climate talks must agree aviation CO_2 cap – UK adviser.* London: Thompson Reuters Foundation Alertnet.

19 Intergovernmental Panel on Climate Change (2007). *Fourth Assessment Report (AR4) of the IPCC.* More than 450 lead authors, 800 contributing authors, and an additional 2,500 reviewing experts from more than 130 countries contributed to AR4.

20 Wang, G. and Eltahir, E.A.B. (2002). Impact of CO_2 concentration changes on the biosphere-atmosphere system of West Africa. *Global Change Biology*, 8, 1169–82.

21 Intergovernmental Panel on Climate Change (2007). Ibid.

22 Burke, M.B., Miguel, E., Satyanath, S., Dykema, J.A., and Lobell, D.A. (2009). Warming increases the risk of civil war in Africa. *Proceedings of the National Academy of Sciences*, 106(49), 20670–74.

23 Anderson, C.A., Bushman, B.J., and Groom, R.W. (1997). Hot years and serious and deadly assault: Empirical tests of the heat hypothesis. *Journal of Personality and Social Psychology*, 73, 1213–23.

24 Hancock, P.A., Ross, J.M., and Szalma, J.L. (2007). A meta-analysis of performance response under thermal stressors. *Human Factors*, 49, 851–77.

25 Burke, M.B., Miguel, E., Satyanath, S., Dykema, J.A., and Lobell, D.A. (2009). Ibid., p. 20673.

26 Bertalanffy, L. von (1969). *General system theory: Foundations, development, applications.* New York: G. Brazilier.

27 Cilliers, P. (1998). *Complexity and postmodernism: Understanding complex systems.* London: Routledge.

28 Cilliers, P. (1998). Ibid.

29 Heylighen, F., Cilliers, P., and Gershenson, C. (2005). *Complexity and philosophy.* Brussels, Belgium: Vrije Universiteit, Evolution, Complexity and Cognition, p. 8.

30 Canadell, J.G. et al. (2007). Contributions to accelerating atmospheric CO_2 growth from economic activity, carbon intensity, and efficiency of natural sinks. *Proceedings of National Academy of Sciences*, 104(47), 18866–70.

31 National Science Foundation (2010, March 4). *Methane releases from Arctic shelf may be much larger and faster than anticipated* (Press Release 10–036). Washington, DC: NSF.

32 Kune, G. (2003). Anthony Eden's bile duct: Portrait of an ailing leader. *ANZ Journal of Surgery*, 73, 341–5. See also Pearson, J. (2003). *Sir Anthony Eden and the Suez Crisis: Reluctant gamble.* Basingstoke: Palgrave Macmillan.

33 Dekker, S.W.A. and Hugh, T.B. (2008). Laparoscopic bile duct injury: Understanding the psychology and heuristics of the error. *ANZ Journal of Surgery*, 78, 1109–114.

34 The West Africa connection: How drug cartels found new routes. *The Times* (of London), February 28, 2009.

35 Kauffman, S. (1993). *The origins of order.* Oxford, UK: Oxford University Press.

36 Nyce, J.M. and Dekker, S.W.A. (2010). IED casualties mask the real problem: it's us. *Small Wars and Insurgencies*, 21(2), 409–413.

37   Byrne, G. (2002). *Flight 427: Anatomy of an air disaster*. New York: Copernicus books.

38   Hutchins, E., Holder, B.E., and Alejandro Pérez, R. (2002). *Culture and flight deck operations* (Prepared for the Boeing Company, Sponsored Research Agreement 22–5003). La Jolla, CA: University of California San Diego.

39   Onderzoeksraad voor de veiligheid (2010). Aircraft Accident Report: *Neergestort tijdens de nadering, Boeing 737–800, nabij Amsterdam Schiphol Airport, 25 februari 2009*. The Hague, NL: OVV.

40   Capra, F. (1975). *The Tao of physics*. London: Wildwood House, p. 139.

41   Capra, F. (1996). *The web of life: A new scientific understanding of living systems*. New York: Anchor Books.

42   Hollnagel, E. (2004). *Barriers and accident prevention*. Aldershot, UK: Ashgate Publishing Co.

43   Johnson, S. (2001). *Emergence: The connected lives of ants, brains, cities and software*. New York: Scribner.

44   Leveson, N. (2006). Ibid.

45   Hollnagel, E. (2004). Ibid.

46   Weick, K.E. (1990). The vulnerable system: An analysis of the Tenerife air disaster. *Journal of Management*, 16, 571–593.

47   Daulerio, A.J. (2010). How one energy company will prevent catastrophic oil spills: Swivel-chair safety. [Online]. Available at: www.deadspin.com [accessed: 17 June 2010].

48   Glanz, J. and Rubin, A.J. (2007). From errand to fatal shot to hail of fire to 17 deaths. *New York Times*, October 3.

49   Risen, J. and Mazzetti, M. (2009). Blackwater guards teamed with C.I.A. on raids. *International Herald Tribune*, December 12–13, pp. 1 and 4.

50   Glanz, J. and Rubin, A.J. (2007, October 3). Ibid.

51   Columbia Accident Investigation (2003, August). *Report Volume 1*. Washington, DC: CAIB, pp. 122–3, 126 and 130.

52   Page, S.E. (2010). *Diversity and complexity*. Princeton, NJ: Princeton University Press.

7
MANAGING THE COMPLEXITY
OF DRIFT

It started out as pure, clear, legitimate deals. And each deal gets a little messier and messier. We started out just taking one hit of cocaine. Next thing you know, we're importing the stuff from Colombia.

Former Enron executive

COMPLEXITY, CONTROL AND INFLUENCE

The traditional view is that organizations are Newtonian–Cartesian machines with components and linkages between them. Accidents get modeled as a sequence of events (actions-reactions) between a trigger and an outcome. But such theories can say nothing about the build-up of latent failures, about a gradual, incremental loosening or loss of control that seems to characterize system accidents and drift into failure.

Remember the basic message of this book. The growth of complexity in society has got ahead of our understanding of how complex systems work and fail. Our technologies have got ahead of our theories. Our theories are still fundamentally reductionist, componential and linear. Our technologies, however, are increasingly complex, emergent and non-linear. Or they get released into environments that make them complex, emergent and non-linear. If we keep seeing complex systems as simple systems – because of the dominant logic and inherited scientific-engineering language of Newton and Descartes – we will keep missing opportunities for better understanding of failure. We will keep missing opportunities to develop fairer responses to failure, and more effective interventions before failure.

One of the patterns through which complex systems spawn accidents is the drift into failure: a slow but steady adaptation of unruly technology to environmental pressures of scarcity and competition. Traditional safety and decision theories are incapable of dealing with drift. They only take snapshots of a system (often an already failed system, for example, layers with holes in them), and cannot bring longitudinal trajectories out in the open. They look for the engine of drift in the wrong actions or decisions of individual components. They make assumptions about rationality and human choice that are impossible to validate in complex systems.

To try to get a better handle on drift, this book has explored the ideas of complexity and systems theory. Can they help us characterize drift into failure in ways that are not componential? Can they shed light on the uncertainty and incompleteness of knowledge of the actors who work inside complex systems? Can they help us recognize the non-linearity and complexity of the intertwined social, technological and organizational processes that make a system descend into failure? Whereas previous ideas about drift and adaptation often remain Newtonian and are "systemic" only because they include more components (see Chapter 5), the ideas of complexity and system science (see Chapter 6) are opening up a new vocabulary to think about non-linear interactions, interdependencies and trajectories toward failure. The problem of drift into failure discussed in this book is not just complex in some vague metaphoric sense, but in the formal sense laid out in the Chapter 6. Drift into failure involves the interaction between diverse, interacting and adaptive entities whose micro-level behaviors produce macro-level patterns, to which they in turn adapt, creating new patterns.

Even with this language, however, and with the generic ideas it gives us for how things might go wrong in complex systems, are we in any better position to predict and prevent failure? If we spend a lot of resources investigating past failures, does that help us at all in forestalling future ones? In a linear Newtonian system, the relationships between causes and consequences can be traced out either forward or backward in time without analytic difficulty. The conditions of a complex system, in contrast, are irreversible. A complex system because, for one, it is never static. Complex systems continually experience change as relationships and connections evolve internally and adapt to their changing environment. Results emerge in ways that cannot be traced back to the behavior of constituent components. The precise set of conditions that gave rise to this emergence is something that can never be exhaustively reconstructed. This means that the predictive power of any retrospective analysis of failure is severely limited.

Decisions in organizations, to the extent that they can be excised and described separate from context at all, are not single beads strung along some linear cause-effect sequence. Complexity argues that they are spawned and suspended in the messy interior of organizational life that influences and buffets and shapes them in a multitude of ways. Many of these ways are hard to trace retrospectively as they do not follow documented organizational protocol but rather depend on unwritten routines, implicit expectations, professional judgments and subtle oral influences on what people deem rational or doable in any given situation.

Reconstructing events in a complex system, then, is nonsensical: the system's characteristics make it impossible. Investigations of past failures thus do not contain much predictive value for a complex system. After all, things rarely happen twice in exactly the same way, since the complex system itself is always in evolution, in flux. Even having a rough idea about how failure arises in complex systems may not help that much, as Pidgeon and O'Leary pointed out:

> To understand how vulnerability to failures and accidents arises, does not automatically confer predictive knowledge to prevent future catastrophes. For in making this complex move one must forsake the familiar ground of accident analysis and disaster development to enter far more contested waters. It is no simple matter to specify how a theory of institutional vulnerability might then be transposed into one of practical resilience. Indeed, we can ask whether our analyses and theories of past accidents and disasters tell us anything useful at all for designing institutions with better future performance, or whether we are merely left with the observation that complex organizations, faced with turbulent environments, will repeatedly fail us in unpredictable ways (and that the only practical advice to risk managers is to stay fully alert to this possibility)?[1]

Designing our way out of drift in complex systems is probably a hopeless task indeed. We cannot design complexity, nor can we design institutions or organizations in ways that ensure that organizations don't become complex. Complexity happens even if we don't want it to happen. That doesn't mean, however, that we are only the helpless victims of complexity. We do play an active role in shaping complexity, even if we might not know it. Speaking the language of complexity can help us find leverage points for playing that role better – or more sensitively, more attuned to the complexity of the system in which we play it.

In a complex system, an action controls almost nothing. But it influences almost everything. Most managers, just like the investigators who pick over the remains of their systems after these have crashed or burned or exploded, have been blind to this contrast. They have operated with a machine model of their world for the longest time, thinking they can control everything, or that some people at some point in time could have controlled everything (and the system is now broken because a lack of control by those individuals). This model is predicated on linear thinking, on symmetry between causes and effects, on predictability. And it is predicated on control.

The prevailing style of management in the West still reflects these notions. It was in part developed and articulated most obviously by Frederick Taylor, whose 1911 book *The Principles of Scientific Management* not only became a classic in the management literature, but whose effects reverberate in what managers see as common sense, as possible, to this day.

Following reductionist logic to the limit, Taylor suggested we should analyse organizations down to its basic component parts, figure out how each of them worked, develop "one best method" for how to get those parts to work and then put it back together again so as to attain the greatest possible efficiency. The workplace, the organization, a collection of humans, it was all no more than a machine, a machine that could ideally be tweaked to operate like clockwork. Ideas hardly get more purely Newtonian than that. The job of the manager was to make sure that the clock ticked without hiccups, that the machine functioned smoothly. This could be done by making assuring the quality of all its parts, and that their interaction was properly lubricated (particularly with top-down orders of what to do). The resulting organizations were very hierarchical. People were simply expected to carry out their minute jobs to a maximum efficiency, and not bother with anything outside of it. No creativity, no unpredictability. It was the embodiment of Weber's maximally rational bureaucracy. Management was about control; decisions were about control.

As the previous chapters have shown, the machine metaphor, or the Newtonian–Cartesian model of organizations, still perfuse styles of management, inspection and policing today. Managerial control and regulatory enforcement are often seen as solutions to (what is seen as) non-compliance and deviance. As said before, there is nothing inherently false or wrong about this approach, or about constructing an organization as something that needs policing and enforcement and control. But by *only* constructing it that way, we obscure other readings. By believing that this is the only way forward, we deny ourselves other perhaps more constructive avenues for making progress on safety.

The commitment that is called for here is to see safety-critical organizations as complex adaptive systems. That we should try to recognize them as such. And that we should see the limits of characterizations that simplify those systems. We should perhaps try to resist the use of languages that dumb these systems down so they can be squeezed into an excel sheet or a bulletized PowerPoint presentation that identifies the less well-functioning components, or propose an intervention that "chops off the head" of an organization whose activities we want to stop.

Reading these systems as complex and adaptive makes us realize that control is perhaps little more than an illusion, at least in some areas and some parts of how the system works. Order is not easy to impose through control from above. In complex systems order arises because of (or emerges from) the interaction of lower-order components and their interaction with their environment. It is not easy to predict in detail how emergent order will look, but it is easy to predict that some kind of order will emerge. Managers should always expect surprises, no matter how simple their goal may seem. The complex system they manage, after all, is nested in larger complex adaptive systems (ultimately the economy). In that non-linear world, predictions never remain valid for long.

Reading the systems we try to operate as complex and adaptive makes us realize that our actions may not control much, but they *do* influence much. Decisions may get dampened along the way, but they may also get amplified, reproduced, copied. Through positive feedback loops, decisions may produce more similar decisions, which produce even more similar decisions later and elsewhere. That is the sort of path-dependency that can help produce drift into failure. It is important to realize that, in complex systems, the effects of local decisions seldom stay local. In a tightly interconnected and interdependent world, decisions have influences way beyond the knowledge or computational capacity of the one who makes them.

This chapter takes up an important feature of complex systems, diversity, and then runs it through the five features of drift – scarcity and competition, small steps, sensitive dependency on initial conditions, unruly technology, and a contributing regulator. Can these features be turned into possible levers for intervention and improvement? Can the complexity of drift be managed? Can we drift into success? It then moves to the final question: that of complexity and accountability. If failure in complex systems is seen as the result of emergence, where the performance of parts cannot be coupled to system-level outcomes, where do we lay responsibility for failure? Can we actually still do so? The story of Enron's drift into failure will be used here to give these questions context. Even though the chapter will conclude with a proposal for a post-Newtonian ethic of failure, this is something that requires more research and consideration that lie beyond the scope of this book.

DIVERSITY AS A SAFETY VALUE

Insights from resilience engineering and complexity science all point to the importance of one thing in managing safety under uncertainty. And that is diversity. Why is diversity so important? Resilience represents the ability to recognize, adapt to and absorb problem disturbances without noticeable or consequential decrements in performance. Diversity is a critical ingredient for resilience, because it gives a system the requisite variety that allows it to respond to disturbances. With diversity, a system has a larger number of perspectives to view a problem with and a larger repertoire of possible responses. Diversity means that routine scripts and learned responses do not get over-rehearsed and over-applied, but that an organization has different ways of dealing with situations and has a rich store of perspectives and narratives to interpret those situations with.

High-reliability organizations have been presented as an example of this. In encouraging diversity, such organizations show a considerable decentralization of decision-making authority about safety issues. This permits rapid and appropriate responses to dangers by the people closest to the problems at hand. Wildavsky called this "decentralized anticipation," which emphasizes the superiority of entrepreneurial efforts to improve safety over centralized and restrictive top-down policies or structures.[2] Such entrepreneurialism is one way to ensure diversity of response. High reliability theory has shown the need for and usefulness of operational discretion at lower levels in the organization. It has found very collegial processes at work, with considerable operational authority resting at very low levels in the organization in applications ranging from nuclear power plants to naval aircraft carriers and air traffic control. Even the lowest ranking individual on the deck of an aircraft carrier has the authority (and indeed the duty) to suspend immediately any take-off or landing that might pose unnecessary risk.[3] As a result, recommendations made by high-reliability theory include a validation of minority opinion and an encouragement of dissent. It also includes an empowerment of lower-ranking employees, such as nurses in hospitals.

Intervention decisions in safety-critical monitored processes are notoriously difficult. These decisions, particularly in escalating circumstances, are extremely hard to optimize, even with experience accrued over time. If the human operator waits too long with diagnosing a problem and gathering evidence that supports an intervention, the decision to intervene may well come too late relative to problem escalation. Degradation has then taken the process beyond any meaningful ability to re-establish process integrity. The human operator can draw few conclusions other than that intervention was too late (but it is seldom clear how much too late). If, in contrast, the human operator intervenes too early, then any evidence that a problem was developing which warranted intervention may disappear as a result of the intervention. The human operator, in other words, is left with no basis to learn about the timing of her or his intervention. How much too early was it? Was the intervention really necessary?

One area in which this dilemma of intervention presents itself acutely is in obstetrics. The intrapartum is surprisingly dangerous. For a variety of anatomical and physiological reasons, even normal labor is hazardous to both parturient and fetus. Uterine contractions impair placental oxygenation as a normal aspect of the intrapartum, something for which the fetus can compensate to some extent through peripheral vasoconstriction, anaerobic metabolism and glycogenolysis, the process of internally generating energy for particularly the brain to survive in conditions of hypoxia.

The problems of supervisory control and intervention in escalating situations (for example, growing fetal distress) present themselves acutely in obstetrics: interventions of various kinds are possible and called for on the basis of different clinical indications and can involve everything up to a decision to deliver the baby by emergency Caesarean section. The difficulty in timing these interventions right is apparent in the increase of even non-emergency C-sections in various countries despite the lack of persuasive clinical indications as to their necessity and postoperative risks to the mother. The growth of a malpractice insurance crisis in a state like New Jersey (where insurance quotes for obstetricians have in some cases gone up to US$200,000 per year) has pushed many Ob-Gyn practitioners to limit their work to gynecology.[4] New Jersey's acrimony in this attests to the difficult and contested nature of obstetric decisions on how and when to intervene in pregnancy and labor and how these decisions get constructed in hindsight.

Adding team members to an obstetric escalating situation does not necessarily increase diversity because of the differential weighting of the voices that get added. We would need some type of diversity index to explore the resilience of various obstetric team make-ups. One candidate type of diversity measure is the Herfindahl index of market concentration, which tries to map how the market shares of different organizations in a particular branch provide competitive diversity or rather monopolistic power. A high index is close to a monopoly, a low index means great diversity. The Herfindahl index H is computed as follows, where s is the decision share of each participant, or the percentage of the voice they have in making an intervention decision.

$$H = \sum_{i=1}^{N} s_i^2$$

Figure 7.1   The Herfindahl Index

If we apply this to an escalating situation in obstetrics escalates, we can imagine the following. Suppose that the situation begins with two midwives, who both have a 45 percent share in any intervention decision that gets made. There is also a lower-ranking nurse present, who has, say, a 10 percent share. Computing the Herfindahl index leads to a diversity index 2*0.452 + 0.12 = 0.415. Now suppose that an experienced resident enters the delivery room as a result of a request by one of the midwives. The resident is appreciated and has been around for a long time, and gets 80 percent of further intervention decision share. The midwives are left with 9 percent each, the nurse with 2 percent. This gives a diversity index of 0.657 as much decision power is now concentrated with the resident. It is closer to a monopoly. If the resident is a novice, however, who announces that he or she does not do c-sections and refuses to engage in other intervention decisions, then the resident's share could drop to say, 15 percent. This leaves the midwives with 40 percent each and the nurse 5 percent. Diversity goes up as a result, with an index of 0.345.

But if the novice resident decides to call an experienced attending (who is well known in the hospital and well respected and therefore has a 90 percent share in intervention decisions), it could bring the index to 0.813, almost entirely monopolistic. Suppose that the resident is experienced, however, and that the attending and the rest of the team have just gone through resource management and non-technical skills training that aims to help downplay hierarchical boundaries. This would leave the attending with 30 percent share in intervention decision, because he or she has learned how to take minority opinion seriously, and empower nurses and midwives. The resident has 20 percent, the midwives 20 percent each, and the nurse 10 percent. H is now 0.22, the greatest diversity yet.

Complex systems can remain resilient if they retain diversity: the emergence of innovative strategies can be enhanced by ensuring diversity. Diversity also begets diversity: with more inputs into problem assessment, more responses get generated, and new approaches can even grow as the combination of those inputs. As the example shows, however, diversity means not only adding team players, but making sure that there is a consideration of the weighting of their voices in any decisions that get made (or, in terms of complexity theory, different thresholds for when they speak up). In cases of drift into failure such *Challenger* and Enron, unequal weighting of decision voices can be said to have played a role, with the president of Morton Thiokol and the president of Enron enjoying a monopoly on key decisions. Knowledgeable participants, for example, the rocket scientists present in the telephone conference on the eve of the *Challenger* launch, have been depicted as having been dealt a very high threshold for speaking up.[5]

Systems that don't have enough diversity will be driven to pure exploitation of what they think already they know. Little else will be explored and nothing new will be learned; existing knowledge will be used to drive through decisions. In business this has been called a take-over by dominant logic, in politics it has been called group think, and in this book it has been laid among the sources of drift into failure. One of the positive feedback loops that starts working with these phenomena is selection. People who adhere to the dominant logic, or who are really good at expressing the priorities and preferences of the organization in how it balances production and risk (like traders in Enron, as we will soon see), will excel and get promoted. This creates, reproduces and legitimates an upper management that believes in the dominant logic, which offers even more incentives for subordinates to adhere to the dominant logic.

There is, of course, need for a balance in the promotion of diversity. Too much diversity can mean that the system will keep on exploring new options and courses of action, and never actually settle on one that exploits what it has already discovered and learned. In time-critical situations (such as in the delivery room), decisions may have to be taken without extensive exploration. Rather, the knowledge that is available right there and then must be exploited even though better alternatives may lie just around the corner in the future.

TURNING THE FIVE FEATURES OF DRIFT INTO LEVERS OF INFLUENCE

Diversity is a key property of complex systems. It is a property that can be harnessed so that complexity is given a role in preventing a drift into failure rather than creating it. Whether the complexity of drift can be managed depends very much on how we can recognize and capitalize on diversity through the five features of drift described in this book. Can we find ways to use these features to our advantage? If we are armed with diversity as a central tactic, can scarcity and competition, small steps, sensitive dependency on initial conditions, unruly technology, and a contributing regulator be turned into levers for intervention and safety improvement? If we can do that, we can perhaps become better at recognizing and stopping drift.

RESOURCE SCARCITY AND COMPETITION

Resource scarcity and competition lead to a chronic need to balance cost pressures with safety. Remember, in a complex system, this means that thousands smaller and larger decisions and trade-offs that get made throughout the system each day can generate a joint preference, a common organizational direction. The decisions also have no apparent local or even knowable global consequences. Production and efficiency get served in the way in which people pursue local goals, yet safety does not get visibly sacrificed. Such a common direction is an emergent property, it is self-organized. It happens without central coordination, without an intelligent designer who says: "This way!" Indeed, even if an organization explicitly commits to be "faster, better and cheaper," as NASA did in the late 1990s, this still doesn't imply that local actors know exactly what that means for them and their daily work. Their actions and the preferences that get expressed through them are not centrally designed and imposed from above.

Resource scarcity and competition is itself a feature of complexity, of course. It is the natural by-product of operating in an environment with interactions and interdependencies across players, where multiple groups are after the same scarce resources and will try to deliver the same service or product faster, better and cheaper. It is impossible to make these features go away other than deciding to no longer compete, to leave the business, to go do something else. An organization cannot typically decide to operate without scarcity or competition (even monopolies often still suffer from resource scarcity in the markets where they operate).

The tensions and paradoxes that are produced by resource scarcity and competition are natural phenomena for complex systems. We shouldn't necessarily try to resolve them, because we probably won't be successful anyway. Complex systems interact with other complex systems, and this leads to tensions and paradoxes that can never be fully reconciled. And there may be benefits to this. In complex social systems, the seemingly opposing forces of competition and cooperation often work together in positive ways – competition within an industry can improve the collective performance of all participants.

But there is another side to this. When an organization is convinced that its business is going faster, better and cheaper, it should get a bit uncomfortable. Locally optimizing processes so that they become faster, better and cheaper probably makes good local sense. But it may not make good global sense. The sum of local optimizations can be global sub-optimization. Or, as Max Weber pointed out more than a century ago: drives to make organizations completely rational will actually produce inescapable irrationalities. You can't optimize or rationalize everything at the same time.

Of course, we shouldn't give up trying to optimize. Lean thinking is believed to have led to gains in efficiency through a removal of unnecessary or redundant steps even in a system as complex as healthcare. But fully optimizing a complex system is both impossible and undesirable. It is impossible because we will never know whether we have actually achieved a totally optimal solution. Even if things get tweaked optimally in a local sense, there will probably be influences of such steps in other parts of the complex system that we might not even be aware of. The cost of optimization in one locality easily gets exported elsewhere – and show up as a constraint that forces a sub-optimal solution there. Extra work may be required, or extra slack may need to be built in elsewhere to accommodate the pressures or demands or constraints imposed by another part of the system.

Fully optimizing a complex system is undesirable because the lack of slack and margin can translate small perturbations into large events. Tight interdependencies or coupling means that there are more time-dependent processes (meaning they can't wait or stand by until attended to), sequences that are invariant (the order of processes cannot be changed) and little slack (for example, things cannot be redone to get it right).[6] It means that small failures can cascade into bigger ones more quickly than if there is margin.

When the system is highly optimized around pressures of scarcity and competition, this is precisely the time to ask whether there is any margin left to absorb or accommodate disruptions from small events. If there isn't, the system is perched in a critical state (where all things are cheap and fast and good), and small perturbations can trigger large events. This means that we should not become so obsessed with small efficiency gains that we push a system toward a critical state.[7] Without slack, there is no possibility to absorb even small disruptions, and small changes can set in motion a cascade of others, triggering a large-scale breakdown.

Diversity of opinion here can be one way to make the system stop and think when it thinks it can least afford to do so. This can be done by having an actor, or group of actors, in the organization who have the authority, the credibility and the courage to say "no" when everybody says "yes." European courts once used to engineer this diversity into the very circle of advisers that royalty relied on for strategic decisions. Knowing that advisers might agree with royal assessments or decisions just to save their own skin or serve their own interests, a jester would be part of the royal retinue, and cast his or her dissent in poetic or musical tones. This form of delivery not only immunized the jester from the expulsion or persecution that would have befallen normal advisers (the dissent and mockery, after all, could be dismissed as a "joke"), it also represented a pointed, memorable way to make the king or queen think twice. When everybody said "yes," the jester could say "no" and get away with it. Shakespeare, as did others, made jesters feature large in some of his plays. In his *Twelfth Night*, the jester was described as "wise enough to play the fool." Their dissent had effect. King James VI of Scotland was known to be very lazy about reading things before signing them. Exploiting this, his jester George Buchanan (1506–1582) tricked him into abdicating in favor of George for 15 consecutive days. James got the point.[8]

SENSITIVE DEPENDENCY AND SMALL STEPS

Because there is no central designer, or any part, that knows the entire complex system, local actors can change their conditions in one of its corners for a very good reason and without any apparent implications. To them, it's simply no big deal, and in fact it may bring immediate gains on some local goal trade-off that people routinely face (for example, getting the system out the door versus checking it once more). Changing these conditions, however, may generate reverberations through the interconnected webs of relationships. Classifying the BP Liberty project in the Beaufort Sea north of Alaska as onshore rather than offshore even before drilling had started (see Chapter 1) was such a change in the initial conditions of the system. The effects of that classification would reverberate in unforeseeable ways through the design, regulation and operation of everything that followed after. The reclassification of foam debris events from flight safety to maintenance issues in the case of Space Shuttle *Columbia* (see Chapter 4) was another such change in conditions. A difference in an initial condition like this can get amplified (or suppressed) as its influences and post-conditions meander through the system.

These changes in condition (a redefinition or reclassification of a risk event, or a particular denotation of the status of a pre-operational system, for example) are not momentous in themselves. Which is why they are relatively easy to achieve, and hard to detect or stop. These are, in fact, often small steps. They are small steps that help the system locally optimize or rationalize a corner of its operations. Between-flight turnaround times of Space Shuttles become quicker if scratches and gouges in the insulating foam are pasted over with a mix of base chemicals rather than repaired, and if the problems that are noted are seen as belonging to maintenance rather than flight safety. Getting to the approval stage with the Minerals Management Service where oil may be pumped is easier for an onshore operation than an offshore one. These are rational, optimizing steps.

Drift into failure is marked by such small steps, or decrementalism. Constant organizational and operational adaptation around goal conflicts, competitive pressure and resource scarcity produces small, step-wise normalizations. Each next step is only a small deviation from the previously accepted norm, and continued operational success is relied upon as a guarantee of future safety. Small steps can become a lever for managing the complexity of drift. There are two advantages to small steps, after all.

One is that small steps happen all the time as complex systems adapt to their environment and produce new interpretations and behaviors as a result. This means that possible opportunities for reflection on practice offer themselves up quite frequently. Why did we just make that decision? Remember, in a complex system any action can influence almost everything. This means that we must promote a solicitude over the possible impacts of small steps. A local optimization may become a global disaster. Ask around about what the step might mean in different corners of the complex system. Go deeper or wider than you may have done before. And then ask some more.

The second advantage is that the steps are small. Calling on people to reflect on smaller steps probably does not generate as much defensive posturing as challenging their more momentous decisions. Retracting a small step, should people conclude that that is necessary, may also not be as expensive. In other words, small steps could mean that there is political and organizational space for critically reflecting on them, and perhaps ultimately space for not taking one of them.

Doing this may require some managerial action, however. Small steps and the normalization they can entail are often no longer be visible or seen as significant by insiders. Outsiders might think about this very differently, and they may represent a resource that managers could capitalize. Having people come in from other operational workplaces does at least two things. It brings fresh perspectives that can help insiders recalibrate what they consider "normal." It also forces insiders to articulate their ideas about running their system, as they engage in activities to train or induct the newcomer. A question such as "why do you guys do that?" can be taken not just as a call for explanation and training, but as an invitation for critical reflection on one's own practices.

This was one of the findings of high-reliability research: the 18-month job rotations on naval aircraft carriers, for example, demanded not only that constant training go on. It also offered the system a diversity of perspectives from people who came onboard after having been at a variety of other places. This was found to be one of the main ingredients for creating high-reliability operations in a risky and safety-critical setting.[9] Rotating personnel from different places (but similar operations) for shorter periods can in principle be done in a variety of worlds. It represents an important lever for assuring the kind of diversity that can make small steps visible and open for discussion.

UNRULY TECHNOLOGY

Unruly technology, which introduces and sustains uncertainties about how and when things may fail, can also be turned into a lever for managing the complexity of drift. Recall how technology can remain unruly. Even though parts or sub-systems of a technology can be modeled exhaustively in isolation, their operation with each other in a dynamic environment generates the unforeseeabilities and uncertainties typical of complexity.

We can actually use this to our advantage. Technology in a complex world, whether it is a Gaussian copula or a 15-storey high external Space Shuttle fuel tank, is never "finished." Declaring a technology operational has much more to do with us, and with our organizational and political constraints, than with the technology and how and where it operates. The entire Space Shuttle program, of course, has shown the perils of declaring a technology operational (and no longer developmental or experimental). That small change of a label carried enormous implications in terms of political, budgetary and societal expectations. And through this, that small change of a label contributed to large events later on: the losses of two Space Shuttles.

To capitalize on unruly technology, rather than be vexed by it, we should invert our perspective. We may traditionally have seen unruly technology as a pest that needs further tweaking before it becomes optimal. This is a Newtonian commitment: we can arrive at an exhaustive description of how the technology works, and we can optimize its functioning (that is, find one best way of deploying it). Complexity theory, in contrast, suggests that we see unruly technology not as a partially finished end-product, but as a tool for discovery. The feature of drift discussed above, small steps, is actually a wonderful property here. Small steps can mean small experiments. Many small steps means many small experiments and opportunities for discovery. These are small experiments without necessarily devastating consequences, but at the same time small experiments with the potential to create hugely important results and insights.

There is a beauty in unruliness. Unruly technology, and its violations of our expectations opens up little windows not just on the workings of a complex system and its environment. It also shows the extent our own calibration. If we are surprised by the effects of technology (NASA's initial surprise about the foam strikes, David Li's surprise and dismay about the enormous popularity of a Gaussian copula he invented), than that generates information both about our environment *and* about us. The questions we should reflect on (even if we may not get clear answers) are two-fold: why did this happen, and why are we surprised that it happened? The first question is about the workings of the complex system and its environment. The second is about our (lack of) understanding of those workings.

Now, separating these two questions of course risks instantiating and reproducing the Newtonian belief that there is a world (of technology-in-context) that is objectively available and apprehensible. Knowledge is nothing more than a mapping from object (technology out there) to subject (our minds). It assumes that our discovery about the workings of unruly technology is not in itself a creative process. It is merely an uncovering of distinctions that were already there and simply waiting to be observed. It rests on the belief that observer and the observed are separable and separated. The idea of our knowledge as an objective representation of a pre-existing order is very Newtonian. In it, the job of an analyst who looks at the data coming from unruly technology is to create representations or constructs that mimic or map the world – their "knowledge." When these copies, or facsimiles, do not match reality, it is due to limitations of the analyst's perception, rationality, or cognitive resources, or due to limitations of methods of observation. More data (and more lines of evidence, cleverer experiments) mean better knowledge: better copies, better facsimiles. This stance represents a kind of aperspectival objectivity. It assumes that we are able to take a "view from nowhere," a value-free, background-free, position-free view that is true. This re-affirms the classical or Newtonian view of nature (an independent world exists to which we as researchers, with proper methods, can have objective access).

But does this work? A central pre-occupation of critical perspectives on science and scientific rationality (as offered by Nietzsche, Weber, Heidegger, Habermas, to name a few) for at least a century-and-a-half is that our (scientific) knowledge is not an objective picture of the world as it is. Isn't what we see and what we can know very much a product of, and constrained by, the knowledge and language we already possess? Complexity science does not see the observed and the observer as separable. The observer is not just the contributor to, but in many cases the creator of, the observed. If there is an objective world of unruly technology out there, then we couldn't know it. As soon as an event has happened (for example, a foam strike), whatever past events can be said to have led up to it, undergo a whole range of transformations. The small steps that were taken by an organization (launching a Space Shuttle in low ambient temperature, for example) can get interpreted as contributing to particular outcomes in the behavior of unruly technology, but can also be constructed as irrelevant (as many believe was done by some of the telcon participants on the eve of the launch of Space Shuttle *Challenger*). Recall how actors or observers in a complex system are intrinsically subjective and uncertain about their environment and future, even if global organization emerges out of their local interactions with each other and the environment. What each of these agents knows, or can know, has little or nothing to do with some objective state of affairs, but is produced locally as a result of those interactions – themselves suspended within a vast and complex and non-linear network of relationships and processes.

The very act of separating important or contributory events from unimportant ones by a few decision-makers, then, is an act of construction, of the creation of a story. It is not the *re*construction of a story that was already there, ready to be uncovered. Any sequence of events or list of contributory or causal factors already smuggles a whole array of selection mechanisms and criteria into the supposed "re"-construction. There is no objective way of doing this – all our choices are affected, more or less implicitly, by the backgrounds, preferences, experiences, biases, beliefs and purposes that we have. "Events" are themselves defined and delimited by the stories with which we configure them, and are impossible to imagine outside this selective, exclusionary, narrative fore-structure.

This, again, is where diversity matters. Diversity means importing different narrative fore-structures, different ways of seeing different things. Diversity means the ability to bring entirely different worlds into being, worlds of observation and experience that would perhaps not have existed within the framework available to the organization itself. The notion that there is not "one" true story of what happened with unruly technology, but rather multiple possible narratives, is an aspect of complexity that can be harnessed. Rather than making contradiction and paradox between different stories go away (by politically repressing them, by not listening to them through reporting schemes or employee empowerment), we could see a great opportunity. The space that is left open between different stories becomes a landscape for exploration and learning – not to find the final, true account, but to see how different interpretations lead to different repertoires of action and countermeasure.

CONTRIBUTION OF THE PROTECTIVE STRUCTURE

The last feature of drift into failure is the contribution of the entire protective structure. This structure can include parts of the organization itself (for example, internal controls or quality assurance programs), but also an external regulator and other forms of oversight. These structures are set up and maintained to ensure safety (or at least ensure regulatory compliance that is in turn believed to help ensure safety).

Protective structures themselves typically consist of complex webs of players and interactions, and are exposed to an environment that influences it with societal expectations, resource constraints, and goal interactions. This affects how it condones, regulates and helps rationalize or even legalizes definitions of "acceptable" system performance. It means that there is often something inescapably paradoxical and corruptible about the role of a protective structure. In principle, the idea of a protective structure is the addition of more diversity to the system. Bringing in outsiders is an obviously good idea – it gets done when consultants come in, but also when a regulator comes in. But if we look at the way in which workers from the federal Minerals Management Service were gradually co-opted by the oil industry they were overseeing, then this is an example of a decline in diversity. Viewpoints between operator and regulator about what was risky and what was okay started to overlap, even while promoting an image of control and diversity.

The problem for a protective structure, of course, is that those outsiders cannot be complete outsiders. They have to have some affinity with the industry or organization they want to understand and influence. They have to have some insider knowledge, in other words. But not so much that they will become acculturated, usurped by the worldview they are supposed to assess. This goes for regulators too. Inspectors have to be outsiders and insiders at the same time. If they are just outsiders, they won't know what to look for and miss subtle signs and weak signals. They will also lack credibility in telling the operator, or their own regulator-superiors, what might need to be done. But if inspectors become insiders only, or if all inspection is basically done by the organization itself, the lack of diversity can lead to a halt in exploration of better ways, safer ways, to do work.

Perhaps the terms most closely associated with protective structures – regulation, compliance, oversight and inspection – are all fundamentally mismatched to complexity. Complex systems cannot in principle be regulated. Regulation, in its bare-bones form, means controlling or maintaining particular parameters so as to keep things operating properly. The idea of regulation is locked in a machine metaphor (and therefore Newtonian assumptions) of how an organization works. This includes not only an image of an organization as a collection of parts and interconnections, but the idea that we can arrive at a complete description of how the system works versus how it is supposed to work. This, of course, is where compliance-based approaches come in. If regulators discover gaps between rules and practice, they may typically try to close those gaps by pressing for practice that is more compliant with the rules.

Complex systems, however, produce behaviors that are more akin to living systems than to machines. Self-organization and emergence, let alone creative evolution, are impossible in a machine. And these behaviors are all very difficult to hold up against a set of rules for how the system is supposed to work. Rules cannot even accommodate creative evolution and emergence, because it would mean that somebody or some agency has designed, in advance, the emergence and creative evolution and self-organization and then condensed it all into rules that reflect some ideal-type against which actual evolving practice can be matched. Emergence, creative evolution and self-organization cannot be designed beforehand. That is the whole point of complexity. Remember how this led, in the case of BP, to an approval of deep-sea drilling on the basis of rules for shallow-water drilling. Practice had evolved ahead of the rules. The regulator, entrapped by its compliance-based approach, could do little more than apply a set of rules it already had in hand (even though these rules had little to do with the emerging practices and unruly technology of deep-sea drilling). This is why the literature on unruly technology posits how even non-compliant practice is rule-following (but those rules are informal and unwritten), and that formal rules follow practice (see Chapters 1 and 3).

If a regulator cannot regulate a complex system, then what can it do? Will a regulator always be caught behind the curves of self-organization and emergence, holding a bag of obsolete rules that came from less evolved systems? Languages of compliance and regulation can perhaps be juxtaposed against those of co-evolution and counter-evolution. Rather than a regulator, complex systems should have a co-evolver/counter-evolver. This must be an organization that has the requisite variety not only to have an idea of the complexity of the operational organization (and thus has to co-evolve with how that organization evolves). It should also have requisite variety to counter-evolve. At least in its theories or models, it should be able to generate alternative outcomes to the small steps that get made by the operational organization.

Safety inspection has an important role to play in complex systems. But inspections can be conducted only on parts or sub-systems. A whole complex system cannot be inspected, only parts or sub-systems can be inspected. After all, if a whole complex system can be inspected, it would mean that the inspector (or group of inspectors) can not only achieve a complete description of the actual system (in which case it wouldn't be complex) but that they also have some ideal description against which they can compare it. These are Newtonian commitments (achieving a complete description of the world and comparing that against a description of an ideal world) that do not work for complexity. Safety inspection, however, should not be about just the parts or their failure rates or probabilities. In itself such findings say very little about the trajectory a complex system may be on, and they can easily be rebutted by operational organizations who can show successful and safe continued operations despite parts being run to failure (see Chapter 1).

This is also why the idea of safety oversight is problematic. Oversight implies a big picture. A big picture, in a sense of a complete description, is impossible to achieve over a complex system. Not only is that computationally intractable, complex systems (as said many times before in this book) evolve and change and adapt the whole time. Nailing its description down at one moment in time means very little for how it will look a next moment. If the big picture of oversight, however, implies a sensitivity to the features of complexity and drift, then it might work. Oversight can try to explore complex system properties such as interconnectedness, interdependence, diversity, and rates of learning and selection that go on inside the complex system. This is perhaps the type of oversight that people inside the operational organization are not capable of because of the locality principle (actors and decision-makers in the organization only see their local interactions). But for a regulator it means learning an entirely new repertoire of languages and countermeasures: from componential, determinist, compliance to co/counter-evolving complexity.

Inspection of parts, then, should relentlessly pursue the possible interconnections with surrounding parts and systems, even those that are external to the sub-system or even organization (like caribou herds outside the Prudhoe bay pipeline). These interconnections can be functional (where parts have a functional or known connection to other parts) but in complex systems the interdependencies are more often non-functional. Caribou have nothing functionally to do with oil transportation, and insulating foam on a fuel tank on lift-off has nothing to do functionally with re-entry of the Space vehicle. The interdependencies are non-functional, and it is unlikely that they can all be made visible if an inspector only looks at existing descriptions of the system (drawings, presentations, documentation, visual and physical inspections). Exploring these interdependencies is a matter of listening to multiple stories, even from those whose responsibilities seem to lie outside of the parts in question. Remember from Chapter 1 that in times of crisis, all correlations go to 1. This means that in times of crisis, these people's responsibilities may well suddenly be affected by a part failure way outside their functional area.

DRIFTING INTO SUCCESS

Rather than setting goals and plotting out the road to get there (or keeping a system on a narrow legally-regulated road), we may want to think about ways to enhance the creativity and the diversity of the operations we run or organize or manage or regulate. Complexity says that we need to see strategy (and, ultimately, safety and accidents) as emergent phenomena, not as the logical endpoints of linear organizational journeys that can be easily retraced afterward or foreseen beforehand. Rather than working on *the* safety strategy, or *the* accident prevention program, we should be creating the pre-conditions that can give rise to innovation, to new emergent orders and new ideas, without necessarily our own careful tending and crafting. In complex systems there are a thousand ways (or more, many more) in which small steps can become big events. No strategy tinkered together by a smart designer (or team of designers) in the organization itself or the regulator can foresee and prevent them all. It is possible, with an approach that promotes creativity and diversity, that a system might even drift into success.

FROM A PAPERCLIP TO A HOUSE

A 26-year-old Montreal man appears to have succeeded in his quest to barter a single, red paper-clip all the way up to a house. It took almost a year and 14 trades, but Kyle MacDonald has been offered a two-storey farmhouse in Kipling, Saskatoon, Canada, for a paid role in a movie.

MacDonald began his quest last summer when he decided he wanted to live in a house. He didn't have a job, so instead of posting a resumé, he looked at a red paper-clip on his desk and decided to trade it on an internet website.

He got a response almost immediately – from a pair of young women in Vancouver who offered to trade him a pen that looks like a fish. MacDonald then bartered the fish pen for a handmade doorknob from a potter in Seattle. In Massachusetts, MacDonald traded the doorknob for a camp stove. He traded the stove to a U.S. marine sergeant in California for a 100-watt generator. In Queens, N.Y., he exchanged the generator for the "instant party kit" – an empty keg and an illuminated Budweiser beer sign.

MacDonald then traded the keg and sign for a Bombardier snowmobile, courtesy of a Montreal radio host. He bartered all the way up to an afternoon with rock star Alice Cooper, a snow globe and finally a paid role in a Corbin Bernsen movie called Donna on Demand. "Now, I'm sure the first question on your mind is, 'Why would Corbin Bernsen trade a role in a film for a snow globe?' MacDonald said. Well, Corbin happens to be arguably one of the biggest snow globe collectors on the planet."

Now, the town of Kipling, Sask., located about two hours east of Regina with a population of 1,100, has offered MacDonald a farmhouse in exchange for the role in the movie. MacDonald and his girlfriend will fly to the town next Wednesday. "We are going to show them the house, give them the keys to the house and give them the key to the town and just have some fun," said Pat Jackson, mayor of Kipling. The town is going to hold a competition for the movie role.

MacDonald said: "There's people all over the world that are saying that they have paper-clips clipped to the top of their computer, or on their desk or on their shirt, and it proves that anything is possible and I think to a certain degree it's true."[10]

Drifting into success is possible because in complex systems the source of emergent phenomena like safety and accidents is not the individual component. These phenomena do not appear as a logical endpoint of a linear organizational journey. They result, rather, from the complex interactions among multiple, diverse, interconnected and interdependent agents who mutually affect each other. This means a couple of things:

o   We need to attend to relationships that can help bring fresh perspectives to the fore, that can help novelty emerge. While we can influence who gets to talk to whom by certain organizational and institutional arrangements, and influence the weighting of their voices when they do get to talk, we should never believe that we can perfectly circumscribe this. In complex systems, there are many more relationships and changes in relationships than we can predict or keep track of. Unforeseen people will talk and influence decisions in ways that lie beyond detection or control.

○   Building the preconditions for diversity, however, is one strong candidate for managing complexity. Greater diversity of agents in a complex system leads to richer emergent patterns. It typically produces greater adaptive capacity. Seeking a diversity of people, cultures, (technical) languages and expertise (and even age and experience) will likely enhance creativity. More stories get told about the things we should be afraid of, more hypotheses may be generated about how to improve the system.

○   Small changes can lead to large events. Changing the language of organizational procedures connected to risk in one way or another (calling something "in-family" or "run-to-failure") can seem to be no big deal. But it can produce reverberations which over time constrain or condition what people elsewhere in the system will see as rational courses of action.

○   Emergence is certain, be we cannot be certain what will emerge. Designing solutions to complex system safety problems is not likely to be successful by itself. Of course, component re-design is always one option (such as the IV ports in the example that started off Chapter 3: they can be redesigned to make certain complex system events less likely). But solutions to safety problems evolve; they don't just follow the path of the designer. The world in which they are supposed to work, after all, is too complex and unpredictable.

COMPLEXITY, DRIFT, AND ACCOUNTABILITY

Remember the question from Chapter 1: Who messed up here? Newtonian logic allows us to answer it relatively easy. But a Newtonian narrative of failure achieves its end only by erasing its true subject: human agency and the way it is configured in a hugely complex network of relationships and interdependencies. The Newtonian identification of a broken part becomes plausible only by obscuring all those interdependencies, only by isolating, mechanizing, or de-humanizing human agency, by making it into a component that zigged but should have zagged. This is both existentially and morally comforting. It is existentially comforting because it allows us to pinpoint a cause for bad things. If we take away the cause or somehow isolate it, we can rest assured that such bad things won't happen again. It is morally comforting because it allows us to place responsibility in the hands of the people whose risk management was deficient: the Newtonian story allows us to hold them accountable.

Since ideas about systemic accident models were first published and popularized, however, system safety has been characterized as an emergent property. Safety is something that cannot be predicted on the basis of the components that make up the system. Accidents, too, have also been characterized as emergent properties of complex systems. They cannot be predicted or explained on the basis of the constituent parts. Rather, they are one emergent result of the constituent components doing their (normal) work. Drifting into failure is possible in an organization where people themselves suffer no noteworthy incidents, in which everything looks normal, and everybody is abiding by their local rules, common solutions, or logics of action.

Recall that emergence means that the behavior of the whole cannot be explained by, and is not mirrored in, the behavior of constituent components. No part needs to be broken for the system to break. Instead, the behavior of the whole is the result – the emergent, cumulative result – of all the local components following their local rules, and of them interacting with each other in innumerous ways, cross-adapting to each others' behavior as they do so. Going from a model of broken components, and the moral and existential satisfaction it may give us, to an understanding of complexity and its much fuzzier, less determined idea of accountability, can be frustrating and difficult. It would almost seem as if complexity gives no room for morality, that it is in itself amoral. Scott Snook concluded as much when he had studied the shoot-down of two U.S. Black Hawk helicopters in 1993. The two helicopters, carrying UN peace keepers, were downed erroneously by two U.S. fighter jets in the no-fly zone over northern Iraq:

> This journey played with my emotions. When I first examined the data, I went in puzzled, angry, and disappointed – puzzled how two highly trained Air Force pilots could make such a deadly mistake; angry at how an entire crew of AWACS controllers could sit by and watch a tragedy develop without taking action; and disappointed at how dysfunctional Task Force OPC must have been to have not better integrated helicopters into its air operations. Each time I went in hot and suspicious. Each time I came out sympathetic and unnerved. If no one did anything wrong; if there were no unexplainable surprises at any level of analysis; if nothing was abnormal from a behavioral and organizational perspective; then what … ?

Snook's impulse to hunt down the broken components (deadly pilot error, controllers sitting by, a dysfunctional Task Force) was tempered by its lack of results. In the end he came out "unnerved," because there was no way he could clearly identify a "cause" that preceded the effect. The most plausible stories of the incident lay outside dominant Newtonian logic.

With an outcome in hand, its (presumed) foreseeability suddenly becomes quite obvious, and it may appear as if a decision in fact *determined* an outcome; that it inevitably and clearly led up to it. In a complex system, the future is uncertain. Knowledge of initial conditions is not enough because the system can develop in all kinds of unforeseeable ways from there on. Also, complete knowledge of all the laws governing the system is unattainable. This is true for the non-linear dynamics of traditional physical systems (for example, the weather) but perhaps even more so for social systems. Social complex systems, composed of individual agents and their may cross-relationships, after all, are capable of internal adaptation as a result of their experiences with each other and with the system's dynamic environment. This can make the possible landscape of outcomes even richer and more complexly patterned. As a result, a complex system only allows us to speculate about probabilities, not certainties.

This changes the ethical implications of decisions, as their eventual outcomes cannot be foreseen. Decision-makers in complex systems are capable only of assessing the probabilities of particular outcomes, something that remains ever shrouded in the vagaries of risk assessment *before*, and always muddled by outcome and hindsight biases *after* some visible system output.

In complexity and system thinking, not only is there no clear line from cause to effect, there is also no obvious symmetry between them as in a Newtonian system. In a complex system, as we have seen, an infinitesimal change in starting conditions can lead to huge differences later on (indeed – having an accident or not having one). This sensitive dependency on initial conditions removes both linearity and proportionality from the relationship between system input and system output. The asymmetry between "cause" and "effect" has implications for the ethical load distribution in the aftermath of complex system failure. Consequences cannot form the basis for an assessment of the gravity of the cause. Trivial, everyday organizational decisions, embedded in masses of similar decisions and subject to special consideration only with the wisdom of hindsight, cannot be meaningfully singled out for purposes of exacting accountability (for example, through criminalization) because their relationship to the eventual outcome is complex, non-linear, and was probably impossible to foresee.

If we adjudicate an operator's understanding of an unfolding situation against our own truth, which includes knowledge of hindsight, we may learn little of value about why people saw what they did, and why taking or not taking action made sense to them. What is unethical or a mistake to one, may have seemed perfectly rational to somebody else at the time (particularly to the one actually doing the work). This should give some pause for thought about what is ethical to do in the aftermath of a drift into failure. Imposing one normative view onto everybody else could easily be seen as unethical, as unjust, as unreasonable. If we take a story that pretty much challenges almost everybody's moral sensibilities, we can see how complexity and accountability might play out. This is the story of the drift into failure of Enron, which ended up as the biggest bankruptcy ever in the United States in 2002.

ENRON'S DRIFT INTO FAILURE

Enron had its humble beginnings in natural gas. The production and distribution of natural gas had always been seen as a poor cousin of the petroleum industry.

In fact, natural gas had long been wasted or discarded as an unwanted by-product of the production of oil. As its usage for heating and energy increased during the 1960s, its production and distribution remained essentially in government hands. Government set production targets and distribution contracts, as well as consumer prices. The industry was totally regulated, and attracted nobody who was interested in dynamic business management or interested in making money. At the time it was quipped that you could get by in the natural gas industry with making one or two decisions a year. With the onset of deregulation in the 1980s, however, the industry started attracting players who were more inclined to make quick deals and pursue large sums of money. Ken Lay, who had worked his way up through the energy industry and ended up in Houston at the head of the natural gas production and distribution company HNG-InterNorth, saw to it that the name was changed to the more manageable Enron, and hired Jeff Skilling, a McKinsey consultant who had helped HNG in the past.

Skilling largely disregarded – indeed had an active distaste for – the messy details involved in executing a plan. And he did not see that as his role, it wasn't the job of the executive. What thrilled him was the intellectual purity of an idea, not the translation of that idea into something implementable. The challenge facing the natural gas industry was pure and simple indeed: the relationship between sellers and buyers was entirely distorted (because of regulation, Ken Lay would add). Gas up to then had been traded under long-term contracts between producers, pipelines, and local utilities, under prices set by the government. The 1980s saw most of the trading diverted to spot markets, where gas changed hands frantically at the end of each month. The uncertainty inherent in this was the basic problem for everybody. Gas had been seen as dull and static (a network of pipes through which it traveled across the country at leisurely speeds was really the major capital investment to be made), but, in a changing environment it actually represented quite unruly technology. A sudden cold spell in the Northeast could cause prices to rise overnight, which would hurt consumers. A wave of warm weather, in contrast, could push prices down again, which would hurt producers and distributors. Even with a surplus of natural gas, big industrial users could never be guaranteed to have enough of a supply from one month to the next. Offering such guarantees was not interesting to gas producers, because at fixed prices, they might not be able to deliver it, or could lose money if they did. As a result, gas was not seen as a reliable fuel.

This is where Skilling's idea of a "Gas Bank" came in. Modeled on how the finance industry traded in money as a commodity, Skilling proposed that gas be traded as well, diminishing the level of risk and uncertainty for all the players involved. And allowing him to make a bunch of money. Producers would contract to sell their gas to Enron at a particular price, and customers would then contract to buy their gas from Enron at a particular price. Enron would function as a sort of bank, capturing the profits between buying gas and selling it. With a balanced portfolio, Enron could not go wrong. Gas producers, however, turned out to be reluctant to sell gas to Enron at fixed prices. They simply kept believing that the price tomorrow would be better.

When Skilling came onboard at Enron, he immediately changed this. Rather then promising a fixed price for gas deliveries in the future, he offered gas producers cash up front. That changed everything. The Gas Bank started to work. Producers flocked to it, attracted by what was essentially a loan made to them by Enron for the production of gas they in some cases still had to find and develop. But it worked. Skilling even went a step further. He proposed how Enron's gas contracts (essentially their promise to buy and sell gas at a particular price) could be traded as well, just like oil futures. Neither the concept of a bank, nor the trading of futures, of course, was very radical. Together, Skilling's initiatives represented small steps that were legitimate and eagerly awaited by many in the industry. But cumulatively, these small and not unprecedented steps amounted to a phase shift. After this, Enron would never be the same.

Enron's contract trading meant that natural gas became unhinged from its natural constraints – the wells and pipes necessary for its production and transportation. Enron now possessed a portfolio of contracts rather than fixed assets, which would allow it to control the resources as it needed to. For Enron, delivering gas wasn't any longer about the physical capability to do so, it was a financial commitment that could be sold, insured, hedged, traded. And it made room for an instrument that hadn't existed there before: the derivative. Reducing gas to its financial terms meant that people could now be sold the right to buy gas at a particular price (just as they had done with stocks for a long time). Derivatives like swaps, options, puts and forwards allowed Enron to start playing with gas as a commodity in the abstract, even though somewhere, somehow, it was still anchored to physical constraints of production capacity, transportation speed, and of course the vicissitudes of weather, energy demand and consumer preferences. Skilling reveled in the complexity of it all, and celebrated the creativity that Enron brought to the table. He assembled a team of like-minded visionaries around him, and in 1990 the New York Mercantile Exchange began to trade gas futures for the first time. The industry had bought into the Enron model.

This is not where the inspiration from stock exchanges ended. Even before Skilling had joined Enron, he had made a very peculiar (but he would argue, visionary) demand to Ken Lay, a demand that was a make-or-break issue for him. His new business, he told Lay, had to use a form of accounting that was very different from the historical-cost accounting everybody else used in the energy industry. Instead, he wanted to use what is known as mark-to-market accounting. With conventional accounting, an organization books the revenues, profits and losses that stem from a contract as the money flows in and out the door. They look back on the quarter, or year, add it all up and see how they fared relative to what they budgeted or projected. With mark-to-market, however, an organization can book the entire estimated value of a contract for years into the future on the day it signs the contract.

This, of course, seems entirely bizarre. It allows you to book projected earnings for years to come as if you've already got the money in your pocket. As if Boeing would be able to book all future profits of its new aircraft, say the 787 Dreamliner, the day the aircraft design is finished, and without ever having built or delivered one. Imagine how such a sudden huge profit would influence Boeing's stock price. Which was actually the point. Skilling had a good reason for his insistence. Mark-to-market accounting, he argued, reflected a much truer picture of the company's financial reality than historical accounting. Booking future profits, he told everyone, shows the true economic value and prowess of a company, and that is what shareholders want to know. That is, indeed, what stock prices should be based on. Wall street firms, who deal in stock, Skilling added, use mark-to-market accounting the whole time. It is normal.

And so, Enron's filings soon began to include the following phrase: "recognized by unrealized income." It was financial shorthand for NASA's "in-family foam damage." It was all normal, within range, not off-the-chart. And it became increasingly normal. Skilling's solution of mark-to-market created value immediately. And its basic tactic spread through the entire company, taking many forms. Mark-to-market, and similar compensation and valuation structures meant that the organization's earnings process was wrapped up the moment the transactions were finalized. The Enron board had no objections; it seemed very efficient. It approved the mark-to-market method in 1991, and then asked the regulator (the Securities and Exchange Commission, or SEC) to do the same.

But earnings that are merely recognized are unrealized indeed. Booking the profits doesn't mean you've got the cash to run the business with. To meet payroll, for example. Or pay the rent. In fact, having booked the profit of a 20-year contract the day the contract is signed, means that that contract needs to be milked for the coming 20 years to actually generate those profits. And you can't book those profits, because you've already done that. In the information given to them, shareholders learn about the profits when the deals are signed. They see good numbers in those unrealized earnings. This actually also makes it very hard for mark-to-market accounting to keep showing continued growth. If the organization wants to project to shareholders that it will grow by, say, an annual 10 percent, it will need entirely new contracts to sign each year that make it so. Such an arrangement, in complexity theory, would be a multiplier: more means more. The aggregation of percentages means that more growth creates demands to show even more growth the next year. And then more again the next. It is an inescapable trap that only works with limitless optimism and no fundamental resource constraints in the market. The treadmill goes faster, and faster, and faster.

Another problem, of course, is that the market price of natural gas isn't that easy to predict, because of the huge complex of factors that influence it. But that actually means that overestimations of price can be made legitimately, precisely because the price is hard to predict and everybody knows that. The SEC was particularly concerned about this particular point: how would Enron estimate the price so far into the future? Enron replied that it would make its estimates based on "known spreads and balanced positions," (which is stock speak, not gas speak), and that they would not be "significantly dependent on subjective elements."[11] The SEC gave in to Enron in 1992. As with all complex systems and their unruly technology, formal rules followed a practice that had already evolved, a practice that was already seemingly operational.

Skilling and his people celebrated. Two weeks later, they wrote to the SEC that the most appropriate period to adopt mark-to-market accounting was not 1992, the year that had just started, but 1991, the year that had just ended. Enron hadn't filed its 1991 statements yet, after all, so why not project a future-projected accounting practice back into the past year? Enron could now report 242 million dollars in earnings for the past year right there. Mark-to-market accounting did wonders for Enron's stock price. Over the years, Enron would extend its mark-to-market accounting to areas where future prices were even harder to estimate. Trading was originally a minor support function inside Enron, a small division, trying to build business from scratch. It didn't stay that way, though. Not many years later, virtually all business activity at Enron would revolve around its giant trading floor. A giant deal helped in this transition. In 1992, Skilling convinced a New York plant developer to build a new power station that ran on gas instead of coal. The future profits that Enron traders marked-to-market were amazing: up to four billion dollars. People started paying attention. Skilling's two-year old "start-up" division inside Enron was now valued at a staggering 650 million dollars.

Meanwhile, in another part of Enron led by Skilling's fiercest rival, Rebecca Mark, a version of mark-to-market accounting cropped up as well, instantiating and reproducing the bullish pre-occupation with growth and cashing in early on signed contracts. The division led by Mark was called Enron International, or EI, and focused in large part on emerging markets. In the first half of the 1990s, Wall Street loved that phrase. It invoked opportunity, growth, unlimited pools of consumers that could be tapped into afresh. Developers in Enron's international division got bonuses on a project-by-project basis for power plants and pipelines and other building projects in far-flung corners of the world. They were always eager to move on to the next deal, rather than following through on the one they just made. There was no financial incentive for it at all. The script got reproduced almost everywhere. In 1994, Enron created a spin-off company that it called Enron Global Power and Pipelines (EPP). The purpose was to purchase developers' assets (power plants, pipelines) to take them off Enron's books. It also freed up capital for Enron to reinvest in new projects. But there was a more important benefit: by selling assets, Enron could realize profits at once, which would help it make its earnings targets promised to Wall Street. It didn't have to wait for these earnings to drift in over years. But who controlled EPP? Enron itself, of course.

By the mid-1990s, Enron had a story that Wall Street loved. It was trading gas contracts aggressively, laying pipelines and nailing breathtaking deals to build power plants in far-flung places. With every passing year, Enron posted record profits. In 1993 it was US$387 million, a year later 453 million and in 1995 520 million. Enron's stock price tripled over about the same period. Executives not only received generous salaries (Ken Lay's alone approached 1 million dollars per year in the mid-1990s, with an annual bonus of just as much). But that was just the beginning. Executives owned sizable amounts of Enron stock (Lay himself owned 3 million shares around this time).

Where was the regulator in all of this? As explained in Chapter 2, the regulator often plays an active role in the drift into failure. Not just by not doing what it is supposed to according to those who pay for regulation, but by actively participating in the construction of risk and its gradual negotiation over time. Enron had its own Risk Assessment and Control department, which Skilling always pointed to when people inquired about the company's risk-management abilities. Skilling had set up RAC, as it was known, as a stand-alone unit because he knew Wall Street and the SEC wanted to see a strong system of internal controls. RAC had to review all Enron deals over half a million dollars, and Skilling boasted that their risk-management was never up for negotiation. Assumptions got questioned, models got tested, price curves checked, portfolios monitored. RAC had resources too, with up to 150 talented people and a 30 million dollar annual budget.

RAC was made a centerpiece of presentations made to Wall Street analysts, investors and credit-rating agencies. The man described by Skilling as Enron's "top cop," Rick Buy, had the power to say "no" to any deal he felt uncomfortable about. Thanks to this arrangement, Enron was able to take on more risk than other companies, because its investors and rating agencies and external regulators let it. Their market presence and internal controls let them get away with taking on more risk than others, Wall Street analysts opined. Enron had become defined by the way it handled risk, it could play loose with deals because its risk management was so tight. But in Enron, nobody ever said no to a deal. Buy didn't have the power, or never deployed it. Even though Buy was seen as the right man to lead a department that rigorously evaluated risk, he was not the man who could stand up and say no to other Enron executives. Skilling knew this, and exploited it maximally. When he eventually left Enron, he told Lay that he should replace Buy with somebody more aggressive, somebody who was stronger.

But there were others with nominal roles in oversight and control. What about the banks that invested in Enron? What about the accountants who signed off on what they did?

> The big Wall Street investment banks, not to mention the nation's giant accounting firms, had a huge vested interest in the kinds of moves Enron was making to create accounting income. Even before the dawning of the 1990s bull market, a new ethos was gradually taking hold in corporate America, according to which anything that wasn't blatantly illegal was therefore okay – no matter how deceptive the practice might be. Creative accountants found clever ways around accounting rules and were rewarded for doing so. Investment bankers invented complex financial structures that they then sold to eager companies, all searching for ways to make their numbers look better. By the end of the decade, things that had once seemed shockingly deceptive, such as securities that looked like equity on the balance sheet but for tax purposes could be treated as debt, now seemed perfectly fine.[12]

Analysts, paid to have an untainted view of a company's performance, had become part of the complex system itself too. Given the relationships that had been strung along the components of this system, they had little choice but to keep a buy rating on Enron, despite any misgivings or suspicions they might have had. Before the 1990s, investment banking and Wall Street research had been separated by a so-called Chinese Wall. Researchers had to be able to make independent stock calls without being concerned about the effect their calls may have on the investments made by the bank. In the 1990s bull market, however, analysts had become instrumental in getting business to their banks. Favorable ratings helped a lot, particularly when companies were about to go public. Promises of favorable research on the company were rewarded with contracts to the bank to make the initial public offering happen. This, too, followed the logic of decrementalism: with these practices becoming more normal with every deal, companies began to expect buy recommendations from analysts. They simply made it a pre-condition for giving their banks any investment business. Enron, with 100 million dollars per year spent on investment banking, had huge leverage in this. Diversity of opinion and perspective disappeared: everybody was conditioned to think and speak the same way. Which, as you might recall from Chapter 1, is how bubbles get formed.

The departure of Enron president and chief operating officer Rich Kinder in 1996 (with Skilling subsequently taking over as president and COO, and later also as chief executive officer) was also a blow to diversity. Some people declared it was one of the saddest days. People with lots of Enron stock unloaded a lot of it right then. Kinder had been dramatically different as president than Skilling would ever be: a good manager and somebody with some understanding of spendthrift. Kinder had offered some guarantee of diversity and resilience. With him gone, the theory went that the Enron inmates themselves started running the asylum. With him gone, diversity plummeted. It can be seen as one tipping point in Enron's drift into failure.

As his business, or the books that were supposed to reflect that business, kept ballooning, Skilling was able to believe that his vision had paid off. And other indicators suggested the same. Trading gas as a commodity, and then trading the securitization of those trades, had become popular. Wildly popular. So popular that he was no longer alone. In the mid 1990s, a host of other players had flooded into the business with their own gas trading desks. So Skilling had to come up with his next big idea. Trading electricity was the first. Trading broadband capacity would be the next. When that idea was announced in 2000, hundreds of security analysts rushed out the room to call their trading desks right away. Enron stock climbed an unbelievable 26 percent in one day. Enron would become a "new economy" company. A better label was impossible. But that was then. By the time bandwidth trading became a reality, the internet bubble was close to bursting, and there was no worse timing for getting involved in the broadband market.

Before all this happened, Enron had become a trading operation in order to keep booking earnings. By the late 1990s a full third of its 18,000 employees was involved in trading and deal making. Most of those were traders who had descended from Skilling's original "start-up." Indeed, it was one more indication of a lack of diversity. Just adding more people to decision problems doesn't necessarily create more diversity. Skilling surrounded himself by people whom he believed he could motivate by one thing only: money. He let people's creativity run wild, he prized independence and individualism. This would have seemed a good investment in diversity: get different, competitive, smart people in to pursue a variety of projects.

The effect on the organization in the end, however, was the opposite. These were people who were very much like Skilling himself. Diversity of opinion, of interpersonal and communicative style, was hardly present. Lack of diversity begat a lack of diversity. With Kinder gone, Skilling had a much easier time appointing the people he wanted throughout Enron. Promotion in organizations is a kind of selection mechanism that usually inhibits diversity, and it operated in Enron with Skilling at the helm too. If an organization selects the best individual performers for promotion, or the best performing contractors for continued work (the trader with the highest earnings of that quarter, or the contractor with the lowest occupational health and safety statistics), it may seem like a great idea. But the result may be that the organization has collected a preponderance of individuals or contractors who have learned that self-interest is the main objective. Loss of diversity led to a loss of diversity, which is often the result of a complex system passing a tipping point. Enron was no more a diverse energy company. It was a trading company, buying and selling securities.

Admitting to Wall Street that Enron was really involved in just buying and selling stuff, however, would have been really difficult. Wall Street knew the inherent risks of trading. It did it itself, after all. So Skilling kept claiming that Enron was a logistics company. It wasn't involved in speculation, it just found smarter ways to deliver gas and electricity from one point to another, ways that were favorable to everybody involved. But the astounding multitude of deals it had struck simply didn't make it so.

> Especially in the latter part of the 1990s, Enron didn't have anywhere near enough cash coming in the door. Eventually, the whole thing took on a life of its own, with an insane logic that no one at the company dared contemplate: to a staggering degree, Enron's "profits" and "cash flow" were the result of the company's own complex dealings with itself.[13]

By that time, Enron had just about securitized everything, from fuel supply contracts to shares of common stock, to partnership interests, and even the profits it had projected to make from its own assets such as powerplant projects in Puerto Rico, Turkey and Italy. In the three years up to 2000, powerplant securitization generated some 366 million dollars in supposed net income (without the plants actually having been built, let alone having them generate any money). Enron went from short-term boost to short-term boost, and it all looked very impressive on paper. Worse, it created a long series of time bombs with ticking debt, debt that was rarely supported by the true underlying value of the asset (such as it was: often there was merely a deal to start a project, not an asset as a result of that project. And then wildly optimistic assumptions were made about the speed and size of its profitability, all to be booked now, before the end of this quarter).

To keep up the appearance of an advanced energy company with plenty of real cash flow, Enron increasingly relied on using the earnings promised to Wall Street as its starting point. It then pushed this target down into the company, telling traders to find the millions that were still missing toward the end of each quarter. They were told to fill the "holes." And interesting strategies were deployed to fill them, strategies that would show the importance of decrementalism, of small steps that would eventually help produce a large event.

What mattered was putting earnings on the books in time for the quarterly results (even, if under the variety of mark-to-market valuations, they hadn't really been "earned" yet – if indeed they ever could). Enron's internal risk management manual legitimated such logic without apology. It proclaimed how the underlying economic realities were essentially irrelevant. It was accounting that mattered, because:

> Reported earnings follow the rules and principles of accounting. The results do not always create measures consistent with underlying economics. However, corporate management's performance is generally measured by accounting income, not underlying economics. Therefore, risk management strategies are directed at accounting, rather than economic, performance.[14]

Those in Enron who had already made deals regularly went back to large existing contracts to see if they could somehow tweak more value out of them. This could be done by renegotiating the terms, or by restructuring the work for the contract. But often it was a matter of looking at the existing contract differently, by making different projections based on what they had learned so far. The contracts, some of them going back five years or more, were reinterpreted to make them look more profitable. A small move in a long-term pricing curve could generate millions in extra profits.[15] Of course, these curves often went so far into the future, that deriving any certainty from their projections was impossible. Eventually, of course, Enron would have to face the reality of those projected (and already booked) earnings, but that never mattered with the immediate, local deadline of the quarterly report. A similar strategy was delaying the reporting of losses. Like any other company, Enron was supposed to write off any deals that hadn't worked out by the end of the quarter. Richard Causey, the Enron accounting chief, would meet with the heads of the groups that had tried to make the deal:[16]

> "At one meeting, an executive recalled, Causey kept coming back to a dead deal and asking: Was it possible the deal was still alive?
> It wasn't, responded the executive.
> 'So there's no chance of it coming back?'
> No.
> 'Is there even a *little* bit of chance of it coming back?' asked Causey. 'Do you want to look at it again?'
> Finally the executive took the hint – and the deal was declared undead. Enron deferred the hit for another quarter. 'You did it once, it smelled bad,' says the executive. 'You did it again, it didn't smell as bad.'[17]

Repetition itself created legitimacy. It normalized the deviant. The smelly gradually became clean. As a result, it was impossible to pinpoint when the first "crime" was committed, when "the line" was crossed:

When, exactly, did Enron cross the line? Even now, after all the congressional hearings, all the investigative journalism, all the reports, lawsuits and indictments, that's an impossible question to answer …

The Enron scandal grew out of a steady accumulation of habits and values and actions that began years before and finally spiraled out of control. When Enron expanded the use of mark-to-market accounting to all sorts of transactions – was that when it first crossed the line? … Or when it created EPP, that "independent" company to which Enron sold stakes in its international assets and posted the resulting gains to its bottom line?

In each case, you could argue that the effect of the move was to disguise, to one degree or another, Enron's underlying economics. But you could also argue that they were perfectly legal, even above board. Didn't all the big trading companies on Wall Street use mark-to-market accounting? Weren't lots of companies moving debt off the balance sheet? Didn't many companies lump onetime gains into recurring earnings? The answer, of course, was yes. Throughout the bull market of the 1990s, moves like these were so commonplace that they were taken for granted, becoming part of the air Wall Street breathed."[18]

The bull market ended in the spring of 2000. Dot-coms were in free fall and many would soon be out of business. In Enron, smart people had become part of a complex system of their own creation, where their next gamble was supposed to cover their last deal. But that no longer worked. As with the Gaussian copula, the interconnections and interdependencies and complexities that had grown to keep pushing the stock price up, now started to work in reverse. There had been no grand, single designer of the mind-boggling complexity and intricacy of Enron's operation. Skilling had been a cheerleader, for sure, and the mind behind some of the grand ideas. But the place had grown on itself. And now, there was no grand, single designer who could stop it from coming crashing down either. The interdependencies and non-linear feedback loops that had multiplied in this complex system now created its own unstoppable dynamic. Stock kept plummeting, and on Sunday December 2, 2001, Enron's lawyers filed the largest bankruptcy case in U.S. history. Skilling had long since resigned, and Lay claimed innocence: he had reportedly been duped by his underlings.

A POST-NEWTONIAN ETHIC FOR FAILURE IN COMPLEX SYSTEMS

In 2006, Skilling was convicted of 19 counts of conspiracy, fraud, lying to auditors and insider trading. He was to serve a 24-year prison sentence in Waseca, Minn. Ken Lay was convicted of 10 counts of fraud and conspiracy in the same trial. Lay, however, died before he could be sentenced. His guilty verdict was erased under a U.S. legal principle that cancels a conviction if the defendant dies before he can appeal.

Was this "correct?" Was it accurate to see Skilling as a criminal who deserved to be punished? It may have given us the idea of closure, of having the storyline clear. Now we know what happened. And the responsible people have been held accountable. But in a complex system, it can be hard to pin down which actions by whom constituted something criminal or immoral. Perhaps we can never rest assured that we *have* a final story, or even a final verdict. The complexities that increasingly ballooned and billowed out of Enron accounting schemes that initially made sense to local players keep befuddling analysts and investigators to this day – no matter how many lawsuits or court cases may be brought. There were too many interconnections, too many interdependencies, too many cross-interactions. And the complex system that spawned them all – that organic, dynamic, living and constantly changing and evolving and adapting thing – is now dead.

Complexity and systems thinking denies us the comfort of one objectively accessible reality that can, as long as we have accurate methods, arbitrate between what is true and what is false. Or between what is right and what is wrong. Rather than producing an accurate story of what happened, complexity invites to pursue multiple authentic stories. Accuracy, after all, denotes preciseness and correctness. It invokes correspondence – between a real, pre-existing world and our image or mirror of that world. The better the mirror or the image, the higher the accuracy of the story. Accuracy is a Newtonian–Cartesian idea and suggests that we can create one definite account based on certainty and methodological precision. Without an accurate story, without a final story, can we definitely determine accountability?

But accuracy cannot be achieved in complex systems since they defy exhaustive description. And whoever does the describing does so from a limited perspective from somewhere in that system. Authentic stories, in contrast, give us some confidence that the one telling the story was *there*, that she or he has made contact with the world and the perspectives of the people in the story, and that she or he was able to delineate and qualify her or his own personal biases and opinions in the process. A story of a complex system is authentic if it succeeds in communicating some of the vitality of everyday life in that system. Perhaps authentic stories do a better job of giving accounts, of holding people accountable, than the putatively "accurate" ones constructed in a court of law or other forensic processes (like a formal accident investigation). The creator of the authentic story, the one giving the account, has to convince the reader that she or he was genuine to that experience, that is, that she or he grasped and understood the system insiders' world as much as possible according to the insiders' constructions of it.[19] That way, we can begin to learn what it meant to live inside a complex system, to be suspended in webs of relationships and interactions, to feel everyday that one's actions could control very little, but would influence almost everything.

In the end, we ourselves do not stand outside of complexity either. We ourselves are inevitably drawn into the complex systems we wish to understand, whose drift into failure we wish to prevent. In the end, we ourselves might have bought Enron, or re-mortgaged our house against an inflated collateral value when money was cheap. We ourselves might have invested in BP stock while filling up our car routinely at their gas stations, and we may have followed the share price's daily gyrations on a mobile phone whose Coltan was mined from endangered Gorilla territory in Kahuzi-Biega[20] and whose parts were assembled in a sub-contacted Taiwanese factory in the Chinese city of Shenzhen whose migrant workers sported one of the highest suicide rates of the country.[21] The problem of accountability in complex systems is no less complex than the complexity of the system itself. If we blame some parts of the system more than others, we put boundaries around those who we want to hold accountable where we choose, not where the complex system says we should. As far as the complex system is concerned, we can always go further, or back, or out more, or in more. Finding a place to put responsibility is up to us, up to our constructions.

In democratic societies, themselves hugely complex systems, we are the ones who help create (and support the enforcement of) laws that put some things in a category considered legal, and other things in a category considered illegal. The boundaries between these categories shift over time. And they are often enormously vague and negotiable. This has even prompted an amendment to the story of accountability after Enron, a story whose criminality once seemed so straightforward. In 2010, the U.S. Supreme Court sharply curtailed prosecutors' use of an anti-fraud law that had been central in convicting Skilling. The law was known as the "honest services" law, and had been criticized by defense lawyers as the last resort of prosecutors in corruption cases that lack the evidence to prove that money changed hands fraudulently. Its constitutionality had been called into question before (hence its appearance in a Supreme Court case) because the law is very vague. It leaves citizens uncertain about whether their behavior can be constituted as criminal or not. U.S. Supreme Court Justice Scalia has criticized the statute, stating that the clause was so poorly defined that it could be the basis for prosecuting "a mayor for using the prestige of his office to get a table at a restaurant without a reservation." In other words, with this law, you never know whether you are violating the law and engaging in a crime or not. Ginsburg, one of the Supreme Court Justices, wrote "because Skilling's misconduct entailed no bribe or kickback, he did not conspire to commit honest-services fraud under our confined construction." Skilling's lawyer predicted that the ruling could lead to a complete overturning of his verdict and an exoneration.[22]

It is impossible to achieve completeness, finality and accuracy in descriptions of complex systems – whether before, during or after their lifetime. This has far-reaching implications for what we might consider ethical in the aftermath of failure:[23]

○   In a complex system, there is no objective way to determine whose view is right and whose view is wrong, since the agents effectively live in different environments. This means that there is never one "true" story of what happened. That people have different accounts of what happened in the aftermath of failure should not be seen as somebody being right and somebody being wrong, or as somebody wanting to dodge or fudge the "truth." In fact, if somebody claims to have the true story of what happened, it turns everybody else into a liar.

○ A complex system should aim for diversity, and respect otherness and difference as values in themselves. Diversity of narratives can be seen as an enormous source of resilience in complex systems (it is when it comes to biodiversity, after all), not as a weakness. The more angles, the more there can be to learn.

○ In a complex system, we should gather as much information on the issue as possible. Of course, complexity makes it impossible to gather "all" the information, or for us to even know how much information we have gathered. But knowing less almost always puts us at a disadvantage when it comes to rendering a verdict about whether certain actions taken inside a complex system are to be seen as right or wrong.

○ In a complex system, we should consider as many of the possible consequences of any judgment in the aftermath of failure, even though it is of course impossible to consider all the consequences. This impossibility does not discharge people of their ethical responsibility to try, particularly not if they are in a position of power; where decisions get made and sustained about the fate of people involved, or about the final word on a story of failure.

○ In a complex system, we should make sure that it is possible to revise any judgment in the wake of failure when as it becomes clear that it has flaws. Of course, the conditions of a complex system are irreversible, which means that even when a judgment is reversed, some of its consequences (psychological, practical) will probably remain irreversible.

We can all work on letting a post-Newtonian ethic of failure emerge if we embrace systems thinking more seriously than we have before. In a post-Newtonian ethic, there is no longer an obvious relationship between the behavior of parts in the system (or their malfunctioning, for example, "human errors") and system-level outcomes. Instead, system-level behaviors emerge from the multitude of relationships, interdependencies and interconnections inside the system, but cannot be reduced to those relationships or interconnections. In a post-Newtonian ethic, we resist looking for the "causes" of failure or success. System-level outcomes have no clearly traceable causes as their relationships to effects are neither simple nor linear.

In such an ethic, we acknowledge that our selection of causes (or events or contributory factors) is always an act of construction. Our extraction of accountability is that too. It is the creation of a story, not the *re*construction of a story that was already there. There is no objective way of constructing a story of what happened – all analytical choices (of what we look at, where we look, who we talk and listen to, how we write up what we learn) are affected by our background, preferences, language, experiences, biases, beliefs and purposes. We can at best hope and aim to produce authentic stories, stories that are true to experience, true to the phenomenology of being there, of being suspended inside complexity.

This means that there are always multiple possible authentic stories, and we should not seek to affirm or legitimate one interpretation (for example, one story of what happened) and let it drown out all others. History, or the natural evolution of the complex system that makes laws and draws lines, might prove us wrong after all. Rather, we might start celebrating multiple dissenting, smaller narratives (including those of lay communities) that can place things in a new language, that can show us the inside corners of our complex systems that we previously had no idea about. This is liberating, hugely informative, and also potentially unsettling. If we relinquish our Newtonian vision of one, accurately describable world and the control, investigations, regulation and management that we perform inside it, then nothing can be taken as merely, obviously, objectively or unconstructedly "true" any longer.

As we tentatively explore a complexity and systems view of failure and, in the limit, its ethical consequences, the differences between an original Newtonian perspective on failure and a systems perspective should become ever clearer. Complexity theory has no answers to who is accountable for drift into failure. Just as the Newtonian view only has oversimplified, extremely limited, and probably unjust answers. What complexity theory allows us to do, however, is dispense with the notion that there *are* easy answers, supposedly within reach of the one with the best method or most objective viewpoint. Complexity allows us to invite more voices into the conversation, and to celebrate the diversity of their contributions. Truth, if there is such a concept, lies in diversity, not singularity.

REFERENCES

1   Turner, B.A. and Pidgeon, N.F. (2000). Man-made Disasters: Why technology and organizations (sometimes) fail. *Safety Science*, 34, 15–30, pp. 17–18.

2   Wildavsky, A.B. (1988). *Searching for safety*. New Brunswick: Transaction Books.

3   Rochlin, G.I., LaPorte, T.R., Roberts, K.H. (1987). The self-designing high reliability organization: Aircraft carrier flight operations at sea. *Naval War College Review*, 76–90.

4   Zaccaria, A. (2002). Malpractice insurance crisis in New Jersey. *Atlantic Highlands Herald*, 7 November.

5   Vaughan, D. (1996). *The Challenger launch decision: Risky technology, culture and deviance at NASA*. Chicago: University of Chicago Press.

6   Perrow. C. (1984). *Normal accidents: Living with high-risk technologies*. New York: Basic Books.

7   Axelrod, R. and Cohen, M.D. (2000). *Harnessing complexity: Organizational implications of a scientific frontier*. New York: Basic Books.

8   Otto, B.K. (2001). *Fools are everywhere: The Court Jester around the world*. Chicago, IL: University of Chicago Press.

9   Rochlin, G.I., LaPorte, T.R., Roberts, K.H. (1987). Ibid.

10   CBC. (2006). *From paperclip to house, in 14 trades*. Ottawa: Canadian Broadcasting Corporation, July 7.

11   McLean, B. and Elkind, P. (2004). *The smartest guys in the room: The amazing rise and scandalous fall of Enron*. New York: Penguin, pp. 41–2.

12   McLean, B. and Elkind, P. (2004). Ibid., pp. 132–3.

13   McLean, B. and Elkind, P. (2004). Ibid., p. 133.

14   McLean, B. and Elkind, P. (2004). Ibid., p. 132.

15   McLean, B. and Elkind, P. (2004). Ibid., p. 128.

16   Calkins, L.B. (2007). Causey heads to prison for role in Enron: Former accounting chief pleaded guilty to fraud. The *Washington Post*, January 3.

17   McLean, B. and Elkind, P. (2004). Ibid., p. 128.

18   McLean, B. and Elkind, P. (2004). Ibid., pp. 132–3.

19   Golden-Biddle, K., and Locke, K. (1996). Appealing work: An investigation of how ethnographic texts convince. *Organization Science*, 4(4), 595–616.

20   United Nations Security Council (1). *Security Council, 4317th and 4318th meeting, condemns illegal exploitation of Democratic Republic of Congo's natural resources* (UN Press Release SC/7057). New York: United Nations.

21   British Broadcasting Corporation. (2010). Foxconn suicides: "Workers feel quite lonely." London: BBC News, 28 May.

22   Associated Press. (2010). *High court reins in prosecutors' use of fraud law*. Washington, DC: AP, June 24,

23   Cilliers, P. (1998). *Complexity and postmodernism: Understanding complex systems*. London: Routledge.

BIBLIOGRAPHY

Althusser, L. (1984). *Essays on ideology*. London: Verso.

Amalberti, R. (2001). The paradoxes of almost totally safe transportation systems. *Safety Science*, 37, 109–26.

Anderson, C.A., Bushman, B.J., and Groom, R.W. (1997). Hot years and serious and deadly assault: Empirical tests of the heat hypothesis. *Journal of Personality and Social Psychology*, 73, 1213–23.

Axelrod, R. and Cohen, M.D. (2000). *Harnessing complexity: Organizational implications of a scientific frontier*. New York: Basic Books.

Bertalanffy, L. von. (1969). *General system theory: Foundations, development, applications*. New York: G. Brazilier.

Billings, C.E. (1996). *Aviation Automation: The Search For A Human-Centered Approach*. Hillsdale, NJ: Lawrence Erlbaum Associates.

Burke, M.B., Miguel, E., Satyanath, S., Dykema, J.A., and Lobell, D.A. (2009). Warming increases the risk of civil war in Africa. *Proceedings of the National Academy of Sciences*, 106(49), 20670–74.

Byrne, G. (2002). *Flight 427: Anatomy of an air disaster*. New York: Copernicus books.

Canadell, J.G. et al. (2007). Contributions to accelerating atmospheric CO_2 growth from economic activity, carbon intensity, and efficiency of natural sinks. *Proceedings of National Academy of Sciences*, 104(47), 18866–70.

Capra, F. (1975). *The Tao of physics*. London: Wildwood House.

Capra, F. (1982). *The turning point*. New York: Simon & Schuster.

Capra, F. (1996). *The web of life: A new scientific understanding of living systems*. New York: Anchor Books.

Catino, M. (2008). A review of literature: Individual blame vs. organizational function logics in accident analysis. *Journal of Contingencies and Crisis Management*, 16(1), 53–62.

Cilliers, P. (1998). *Complexity and postmodernism: Understanding complex systems*. London: Routledge.

Cohen, M.D., J.G. March, et al. (1988). A garbage can model of organizational choice. In J.G. March (ed.). *Decisions and organizations*, pp. 294–334. Oxford: Blackwell.

Collins, F.S. (2006). *The language of God*. New York: Free Press.

Columbia Accident Investigation Board. (2003, August). *Report Volume 1*. Washington, DC: CAIB.

Dekker, S.W.A. (2005). *Ten questions about human error: A new view of human factors and system safety*. Mahwah, NJ: Lawrence Erlbaum Associates.

Dekker, S.W.A. (2007). *Just culture: Balancing accountability and safety*. Aldershot, UK: Ashgate Publishing Co.

Dekker, S.W.A. and Hugh, T.B. (2008). Laparoscopic bile duct injury: Understanding the psychology and heuristics of the error. *ANZ Journal of Surgery*, 78, 1109–1114.

Dörner, D. (1989). *The logic of failure: Recognizing and avoiding error in complex situations*. Cambridge, MA: Perseus Books.

Feynman, R.P. (1988). *"What do you care what other people think?": Further adventures of a curious character*. New York: Norton.

Fischhoff, B. (1975). Hindsight is not foresight: The effect of outcome knowledge on judgment under uncertainty. *Journal of Experimental Psychology: Human Perception and Performance*, 1(3), 288–303.

Foucault, M. (1980). Truth and power. In C. Gordon (ed.). *Power/Knowledge*, pp. 80–105. Brighton: Harvester.

Freud, S. (1950). Project for a scientific psychology. In *The standard edition of the complete psychological works of Sigmund Freud, vol. I*. London: Hogarth Press, p. 295.

GAIN. (2004). *Roadmap to a just culture: Enhancing the safety environment*. Global Aviation Information Network (Group E: Flight Ops/ATC Ops Safety Information Sharing Working Group).

Gawande, A. (2002). *Complications: A surgeon's notes on an imperfect science*. New York City: Picador.

Giddens, A. (1991). *Modernity and self-identity: Self and society in the Late Modern age*. London: Polity Press.

Golden-Biddle, K. and Locke, K. (1996). Appealing work: An investigation of how ethnographic texts convince. *Organization Science*, 4(4), 595–616.

Gould, S.J. (1987). Opening remarks of the conference on evolutionary progress at Chicago's Field Museum. Quoted in Lewin, R. (1999). *Complexity: Life at the edge of chaos*. Second Edition. Chicago, IL: University of Chicago Press. p. 139.

Green, J. (2003). The ultimate challenge for risk technologies: Controlling the accidental. In J. Summerton and B. Berner (eds.). *Constructing risk and safety in technological practice*. London: Routledge.

Hancock, P.A., Ross, J.M., and Szalma, J.L. (2007). A meta-analysis of performance response under thermal stressors. *Human Factors*, 49, 851–77.

Heylighen, F. (1989). Causality as distinction conservation: A theory of predictability, reversibility and time order. *Cybernetics and Systems*, 20, 361–84.

Heylighen, F., Cilliers, P., and Gershenson, C. (1995). *Complexity and philosophy*. Brussels: Vrije Universiteit.

Heylighen, F., Cilliers, P., and Gershenson, C. (2005). *Complexity and philosophy*. Vrije Universiteit Brussel: Evolution, Complexity and Cognition.

Holden, R. (2009). People or Systems? To blame is human. The fix is to engineer. *Professional Safety*, 12, 34–41.

Hollnagel, E. (2004). *Barriers and accident prevention*. Aldershot, UK: Ashgate Publishing Co.

Hollnagel, E., Woods, D.D., and Leveson, N.G. (eds.) (2006). *Resilience engineering: Concepts and precepts*. Aldershot, UK: Ashgate Publishing Co.

Hutchins, E., Holder, B., and Pérez, R.A. (2002). *Culture and flight deck operations (Research Agreement 22–5003)*. San Diego: University of California, UCSD.

Jensen, C. (1996). *No downlink: A dramatic narrative about the Challenger accident and our time*. New York: Farrar, Strauss, Giroux.

Johnson, S. (2001). *Emergence: The connected lives of ants, brains, cities and software*. New York: Scribner.

Kauffman, S. (1993). *The origins of order*. Oxford, UK: Oxford University Press.

Kune, G. (2003). Anthony Eden's bile duct: Portrait of an ailing leader. *ANZ Journal of Surgery*, 73, 341–5.

Langewiesche, W. (1998). *Inside the sky: A meditation on flight*. New York, Pantheon Books.

Lanir, Z. (1986). *Fundamental surprise*. Eugene, OR: Decision Research.

LaPorte, T.R. and Consolini, P.M. (1991). Working in Practice but Not in Theory: Theoretical Challenges of 'High-Reliability Organizations.' *Journal of Public Administration Research and Theory: J-PART* 1(1): 19–48.

Laszlo, E. (1996). *The systems view of the world: A holistic vision for our time*. Cresskill, NJ: Hampton Press.

Leveson, N.G. (2002). *A new approach to system safety engineering*. Cambridge, MA: Aeronautics and Astronautics, Massachusetts Institute of Technology.

Leveson, N.G. (2006). *System Safety Engineering: Back to the future*. Cambridge, MA: Aeronautics and Astronautics, Massachusetts Institute of Technology.

Leveson, N.G. and Turner, C.S. (1993). An investigation of the Therac-25 accidents. *Computer*, 26(7), 18–41.

Leveson, N., Cutcher-Gershenfeld, J., Barrett, B., Brown, A., Carroll, J., Dulac, N., Fraile, L., and Marais, K. (2004). Effectively Addressing NASA's Organizational and Safety Culture: Insights from Systems Safety and Engineering Systems. *Paper presented at the Engineering Systems Division Symposium, MIT*, Cambridge, MA, March 29–31.

Leveson, N., Daouk, M., et al. (2003). *Applying STAMP in accident analysis*. Cambridge, MA, Engineering Systems Division, Massachusetts Institute of Technology.

Lewin, R. (1999). *Complexity: Life at the edge of chaos, second edition*. Chicago, IL: University of Chicago Press.

Mackall, D.A. (1988). *Development and Flight Test Experiences with a Flight-Critical Digital Control System* (NASA Technical Paper 2857). Lancaster, CA: National Aeronautics and Space Administration Dryden Flight Research Facility.

Marone, J.G. and Woodhouse, E.J. (1986). *Avoiding catastrophe: Strategies for regulating risky technologies*. Berkeley, CA: University of California Press.

Martin, J. (1985). Can organizational culture be managed? In Frost, P.J. (ed.). *Organizational Culture*. Thousand Oaks, CA: Sage Publications.

McDonald, N., Corrigan, S., and Ward, M. (2002, June). *Well-intentioned people in dysfunctional systems*. Keynote paper presented at the 5th Workshop on Human error, Safety and Systems development, Newcastle, Australia.

McLean, B. and Elkind, P. (2004). *The smartest guys in the room: The amazing rise and scandalous fall of Enron*. New York: Penguin.

National Science Foundation. (2010, March 4). *Methane releases from Arctic shelf may be much larger and faster than anticipated* (Press Release 10–036). Washington, DC: NSF.

National Transportation Safety Board. (2000). *Factual Report: Aviation (DCA00MA023), Douglas MD-83, N963AS, Port Hueneme, CA, 31 January 2000.* Washington, DC: NTSB.

Nyce, J.M. and Dekker, S.W.A. (2010). IED casualties mask the real problem: it's us. *Small Wars and Insurgencies*, 21(2), 409–413.

Otto, B.K. (2001). *Fools are everywhere: The Court Jester around the world.* Chicago, IL: University of Chicago Press.

Outram, D. (2005) *The Enlightenment* (Second Edition: New approaches to European history). Cambridge, UK: Cambridge University Press.

Page, S.E. (2008). Uncertainty, difficulty and complexity. *Journal of Theoretical Politics*, 20(2), 115–49.

Page, S.E. (2010). *Diversity and complexity.* Princeton, NJ: Princeton University Press.

Pearson, J. (2003). *Sir Anthony Eden and the Suez Crisis: Reluctant gamble.* Basingstoke: Palgrave Macmillan.

Peat, F.D. (2002). *From certainty to uncertainty. The story of science and ideas in the twentieth century.* Washington, DC: Joseph Henry Press.

Pellegrino, E.D. (2004). Prevention of medical error: Where professional and organizational ethics meet. In Sharpe, V.A. (ed.). *Accountability: Patient safety and policy reform.* Washington, DC: Georgetown University Press, pp. 83–98.

Perrow, C. (1984). *Normal accidents: Living with high-risk technologies.* New York: Basic Books.

Poole, R.W. Jr., and Butler, V. (1999). Airline deregulation: The unfinished revolution. *Regulation*, 22(1), 8.

Rasmussen, J. (1997). Risk management in a dynamic society: A modeling problem. *Safety Science*, 27(2/3), 183–213.

Rasmussen, J., and Svedung, I. (2000). *Proactive risk management in a dynamic society.* Karlstad, Sweden: Swedish Rescue Services Agency.

Reason, J.T. (1990). *Human error.* Cambridge, UK: Cambridge University Press.

Rochlin, G.I. (1999). Safe operation as a social construct. *Ergonomics*, 42(11), 1549–60.

Rochlin, G.I., LaPorte, T.R., et al. (1987). The self-designing high reliability organization: Aircraft carrier flight operations at sea. *Naval War College Review*, 76–90.

Rose, B.W. (1994). *Fatal dose: Radiation deaths linked to AECL computer errors.* Montreal, QC: Canadian Coalition for Nuclear Responsibility.

Sagan, S.D. (1993). *The limits of safety: Organizations, accidents and nuclear weapons.* Princeton, NJ: Princeton University Press.

Smetzer, J., Baker, C., Byrne, F., Cohen, M.R. (2010). Shaping systems for better behavioral choices: Lessons learned from a fatal medication error. *The Joint Commission Journal on Quality and Patient Safety*, 36(4), 152–63.

Snook, S.A. (2000). *Friendly fire: The accidental shootdown of US Black Hawks over Northern Iraq.* Princeton, NJ: Princeton University Press.

Stapp, H.P. (1971). S-matrix interpretation of quantum theory. *Physical Review D*, 3.

Starbuck, W.H. and Milliken, F.J. (1988). Challenger: Fine-Tuning the Odds Until Something Breaks. *The Journal of Management Studies*, 25(4), 319–41.

Stech, F.J. (1979). *Political and Military Intention Estimation. Report N00014–78–0727.* Bethesda, MD: US Office of Naval Research, Mathtech Inc.

Tingvall, C. and Lie, A. (2010). The concept of responsibility in road traffic (Ansvarsbegreppet i vägtrafiken). Paper presented at *Transportforum*, Linköping, Sweden, 13–14 January.

Turner, B.A. (1978). *Man-made Disasters.* London: Wykeham.

Turner, B.A. (1995). A personal trajectory through organization studies. *Research in the Sociology of Organizations*, 13, 275–301.

Turner, B.A. and Pidgeon, N.F. (1997). *Man-made disasters* (Second Edition). Oxford: Butterworth Heinemann.

Turner, B.A. and Pidgeon, N.F. (2000). Man-made disasters: Why technology and organizations (sometimes) fail. *Safety Science*, 34, 15–30.

Vaughan, D. (1996). *The Challenger launch decision: Risky technology, culture and deviance at NASA.* Chicago: University of Chicago Press.

Vaughan, D. (1999). The dark side of organizations: Mistake, misconduct, and disaster. *Annual Review of Sociology*, 25, 271–305.

Vaughan, D. (2005). System effects: On slippery slopes, repeating negative patterns, and learning from mistake? In W.H. Starbuck and M. Farjoun (eds). *Organization at the limit: Lessons from the Columbia disaster*, pp. 41–59. Malden, MA: Blackwell Publishing.

Wang, G. and Eltahir, E.A.B. (2002). Impact of CO_2 concentration changes on the biosphere-atmosphere system of West Africa. *Global Change Biology*, 8, 1169–82.

Weick, K. (1995). *Sensemaking in organizations.* London: Sage.

Weick, K.E. (1990). The vulnerable system: An analysis of the Tenerife air disaster. *Journal of Management*, 16, 571–93.

Weick, K.E. (1993). The collapse of sensemaking in organizations: The Mann-gulch disaster. *Administrative Science Quarterly*, 38(4), 628–52.

Weick, K.E. and Sutcliffe, K.M. (2007). *Managing the unexpected: Resilient performance in an age of uncertainty.* San Francisco: Jossey-Bass.

Weingart, P. (1991). Large technical systems, real life experiments, and the legitimation trap of technology assessment: The contribution of science and technology to constituting risk perception. In T.R. LaPorte (ed.). *Social responses to large technial systems: Control or anticipation*, pp. 8–9. Amsterdam: Kluwer.

Wijaya, M. (2002). *Architecture of Bali: A source book of traditional and modern forms.* Honolulu: University of Hawai'i Press.

Wildavsky, A.B. (1988). *Searching for safety.* New Brunswick: Transaction Books.

Woolfson, C. and Beck, M. (eds). (2004). *Corporate responsibility failures in the oil industry.* Amityville, NY: Baywood Publishing Company, Inc.

Wynne, B. (1988). Unruly technology: Practical rules, impractical discourses and public understanding. *Social Studies of Science*, 18, 147–67.

INDEX